North Florida Reefs

Jacksonville and St. Augustine Fishing, Diving & Marine Life

JOSEPH KISTEL

KISTEL MEDIA

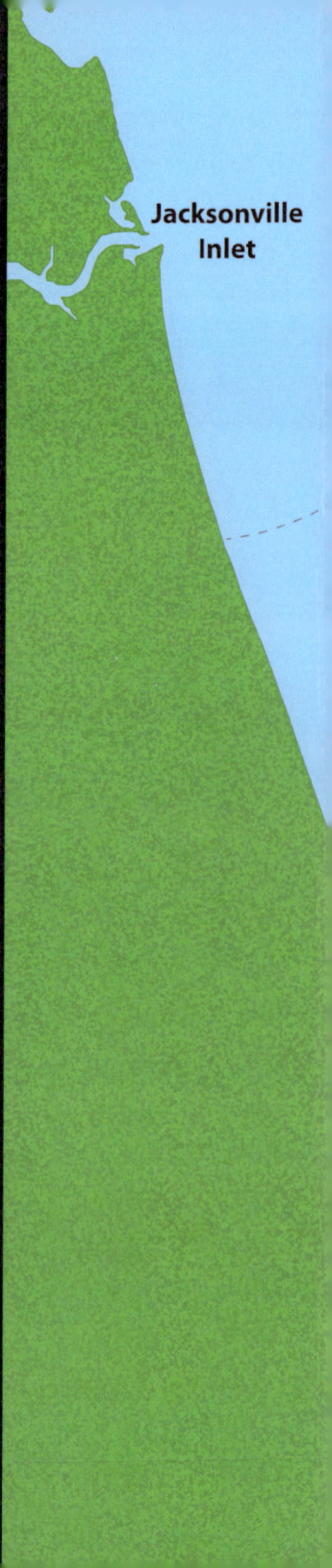

Published by:
Kistel Media
10752 Deerwood Park Blvd. South
Suite 100
Jacksonville, FL 32256
www.KistelMedia.com

ISBN: 978-0-9891719-0-8

Cover picture: Blue Anglefish

Title page picture: Nurse Shark

Jacksonville Inlet

St. Augustine Inlet

10 miles

20 miles

30 miles

Acknowledgments

This book is the result of several years' efforts and many underwater hours. I would like to mention those below who were influential in the collection of content within these pages.

Ed Kalakauskis - My good friend and trusty dive partner who was underwater with me nearly every dive in which the images within were captured. Ed is one of the pioneers of the artificial reef program in northeast Florida and participated in many of the deployments of the reefs featured. He was not only an invaluable information source but also someone I could always count on.

Larry Davis - Larry is a close friend who also participated in several dives. When Larry was not in the water capturing imagery he was always on standby with equipment support. From camera equipment to dive gear, Larry was always selflessly eager to provide whatever was requested.

Offshore Transportation:

Steve Blalock – Owner and captain of the Tickle Stick, Steve provided transportation to nearly all of the St. Augustine reefs featured in "North Florida Reefs." As a veteran of the St. Augustine reef program, Steve was also a wealth of knowledge.

Jack Leone – St. Augustine offshore transportation
John Perkner – Jacksonville offshore transportation
Dan Lindley, "Offshore Dive Charters" – Offshore transportation
Steve Park, "Native Diver II" – Offshore transportation

Reef Habitat Supporting Organizations:

TISIRI, JOSFC, ACGFA, JRRT, City of Jacksonville, City of St. Augustine, Flagler County, and the **FWC** are all organizations that have played significant roles in the reef creation programs of North Florida.

IT and Software Support:

Nathan Tucei of Nassau WebDesign

Facilities Support:

Office Suites Plus

Image Credits:

Larry Davis
The Spike: pictures 2,3, and 6; Bob Engle: pictures 1 and 4; BR Ferry: pictures 1 and 2; Casablanca: pictures 1 and 6
Joseph Kistel
All other images in this publication

About the Author

Joseph Kistel, aka "Scuba Joe," is known for his offshore adventures, underwater discoveries, and marine conservation efforts. He and his underwater imagery have been featured on the Discovery Channel, CNN, Fox, NBC, and ABC.

Kistel is passionate about sharing the amazing opportunities our world's oceans provide, and this book is a good example of that. North Florida is home to several spectacular marine reef systems that unfortunately many people are not aware of. This book was created to highlight some of these reef systems and the complexity of marine life they support. Readers are provided images of actual reef sites and their inhabitants, as well as geographical location information. This publication not only shows readers the reefs, but also how to locate and travel to them.

Kistel is a conservationist at heart and encourages visitation and usage of marine habitats. He believes awareness of these ecosystems is critical to the success of how well society protects and maintains marine resources. Experiencing a reef first hand, via a fishing or scuba diving trip, is arguably the most effective way for an individual to become aware of the significance of marine habitats. Readers are encouraged to visit and utilize these offshore destinations, but to do so in a respectable manner. Keep in mind that like any natural resource, reef ecosystems can be easily damaged from over exploitation by us humans.

Kistel currently serves as director for the marine conservation organization TISIRI and is the head producer at Kistel Media.

North Florida Reefs

Similar to onshore facilities like amusement parks or entertainment venues, offshore reefs are destinations people enjoy visiting for recreation. Reefs are considered attractions because of their abundance of marine life and the opportunities they bring to fishermen, scuba divers, and wildlife enthusiasts. Fortunately, the offshore waters of northeast Florida are scattered with productive reef habitats.

"North Florida Reefs" is intended to serve as a hand-held guide to offshore habitats and marine life found off Jacksonville and St. Augustine. The publication not only provides GPS coordinates of locations, but also includes actual imagery of the reefs and their inhabitants. A photo guide and description of popular fish and invertebrate species encountered at these reefs sites is also included.

Using North Florida Reefs

This book is primarily split into three sections: Jacksonville reefs, St. Augustine reefs, and marine life. At the beginning of each reef's section is a table and map corresponding to the offshore waters of that particular region. The maps include circled numbers representing reef sites in their approximate geographical location. The reefs are numbered in order from their closeness to the corresponding inlet, and these numbers are referred to as the "Reef Map Number" in the tables. Simply locate the "Reef Map Number" on the table to find the name of the reef and the corresponding pages associated with the site. Turn to the pages to find GPS coordinates, a description, and imagery captured at each particular reef.

The marine life section is organized alphabetically based on the common name of the organisms. It is encouraged that readers look and read through every animal featured as some species may be considered quite unusual.

Pages 8 and 9 include an easy to follow contents' breakdown allowing readers to quickly locate relevant subject matter.

Noteworthy

Reefs 1 - 22 are considered Jacksonville reefs while reefs 23 - 38 are considered St. Augustine reefs. Additionally, reef numbers 14 and 15 are nearly equal distance from both Jacksonville and St. Augustine inlets but will be considered Jacksonville reefs in this publication as these reef deployments were coordinated through the City of Jacksonville. Furthermore, reef numbers 32 and 33 are the result of Flagler County reef efforts but are included in the St. Augustine section of this book because of their proximity to the St. Augustine Inlet.

This guide is not meant to serve as a primary source of navigation. Although the GPS coordinates provided have proven accurate in our experience, sometimes variation exists in exact coordinate location based on the GPS unit utilized and current environmental conditions. It is recommended that a good sonar unit is used in conjunction with a high quality GPS to locate sea floor structure.

The dotted arc lines on the maps and the heading-distance information on the reef pages correspond to reference points from the inlets. More specifically, the reference point from the Jacksonville Inlet is where the jetties terminate into the Atlantic, while the reference point for the St. Augustine Inlet is where the inlet meets the Atlantic. Distances and headings provided should be considered approximations, as they were determined by plotting the GPS coordinates on a map and interpreting the distance and heading based on the map scale.

Reef Site Contents

Jacksonville

St. Augustine

Marine Life Contents

Look for these amazing animals

JACKSONVILLE

Jacksonville Florida Reefs

Reef Map #	Reef Name	Page #
1	MR East Barge	14
2	Culverts and Barge	16
3	PG Barge	18
4	Natural Ledge 2	20
5	Natural Ledge 1	22
6	Banana and Pogey	24
7	Gulf America Wreck	26
8	Tugboat #1	28
9	School Bus Barge	30
10	Gibbs Dry Dock	32
11	Press Box Ledge Reef	34
12	Press Box	36
13	CH Tug	38
14	FF Concrete 2011	40
15	FF Concrete 2009	42
16	HG Ledge	44
17	Coppedge Tug	46
18	Coppedge Culverts	48
19	The Spike	50
20	Bob Engle	52
21	BR Ferry	54
22	Casablanca	56

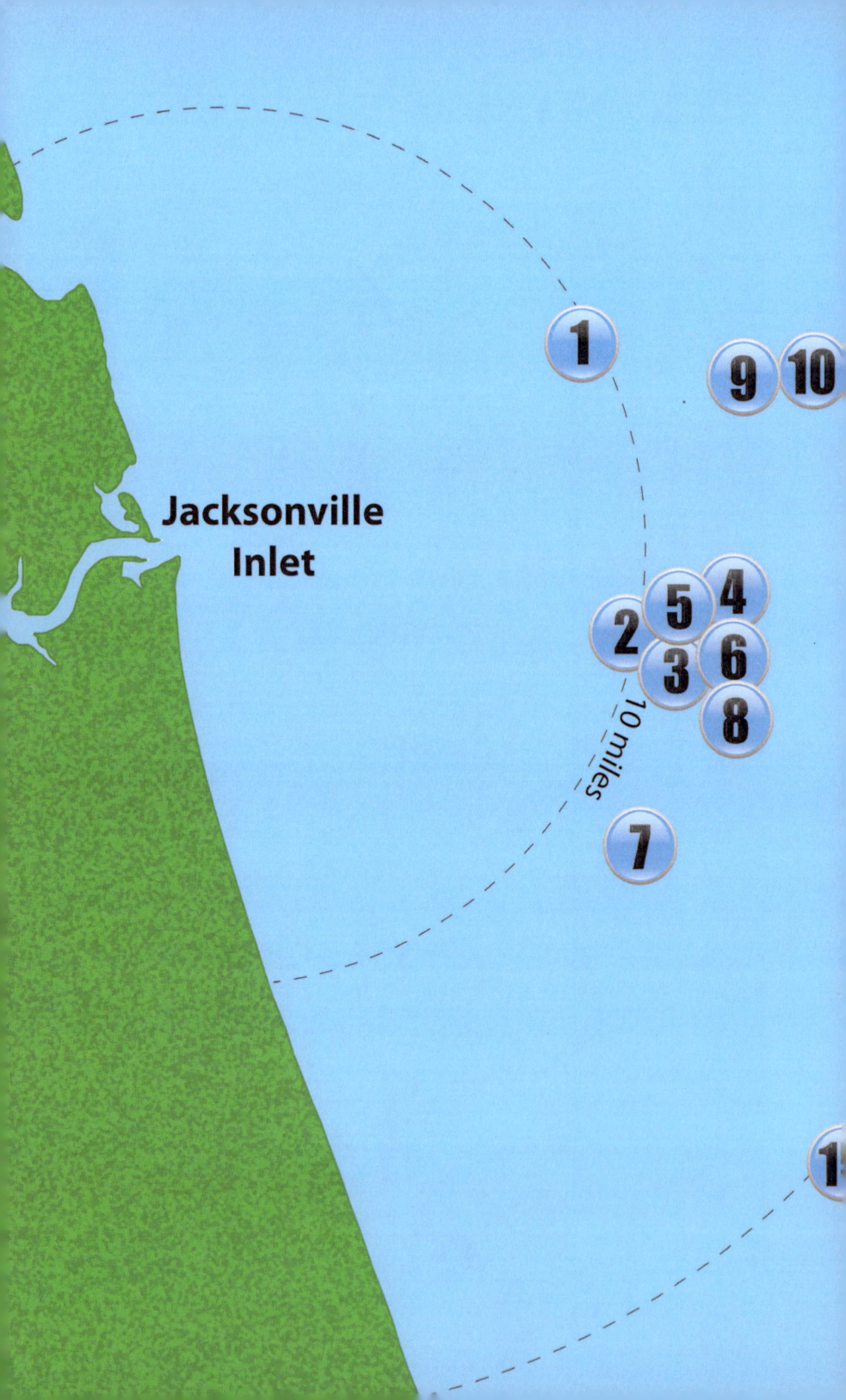

Jacksonville
Inlet

1

9 10

2 5 4
3 6
8

10 miles

7

1

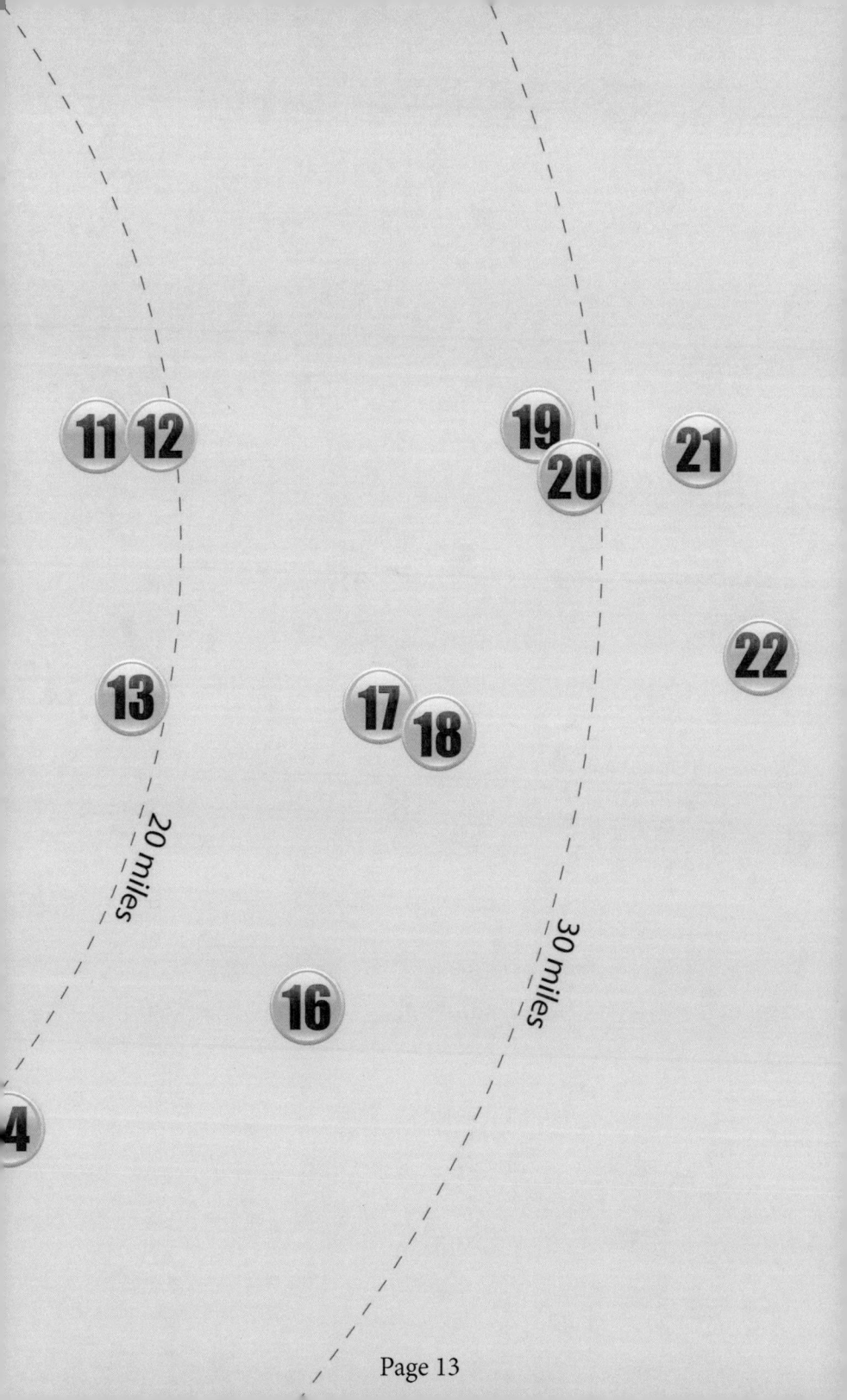

11 12

19
20 21

22

13

17
18

20 miles

30 miles

16

4

1) MR East Barge

GPS Coordinates: 30° 26.689'
                                        81° 13.269'

Depth: 70 feet

Structure: Metal Barge

Distance and Heading
    Jacksonville Inlet: 9.77 miles @ 070.93°
    St Augustine Inlet: 36.64 miles @ 005.35°

Description:

The East Barge Reef is a simple rectangular barge with a complex assortment of marine life. The entire metal structure is encrusted with sponges, corals, and a plethora of other invertebrates. Because of this, an ecosystem exists supporting all tiers of the marine food chain. Being both close to the inlet and relatively shallow, it makes for an ideal destination for open water divers. There are plenty of colorful critters around for the underwater photographer. For the fishermen out there, keep your eyes peeled for flounder in the sand around the perimeter of the barge.

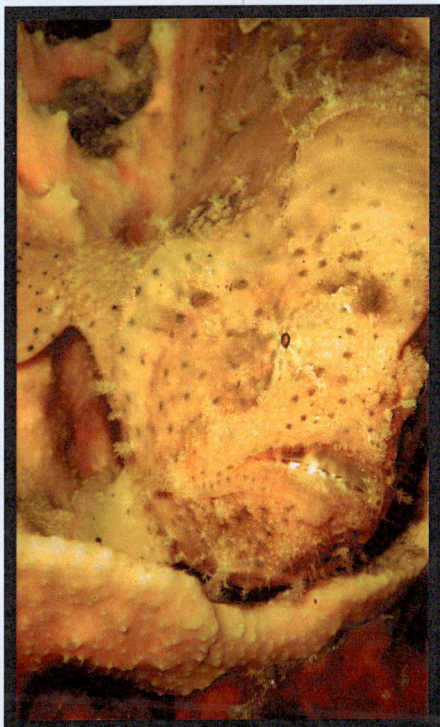

2) Culverts and Barge

GPS Coordinates: 30° 20.846'
                                    81° 12.829'

Depth: 70 feet

Structure: Culverts and Barge

Distance and Heading
  Jacksonville Inlet: 10.34 miles @ 110.03°
  St Augustine Inlet: 30.31 miles @ 007.69°

Description:

This reef is quite diverse as it consists of concrete culvert structures, metal fuel tanks, and a metal barge. The concrete and metal tanks were deployed in the early 1980's and the barge was added at a later date. The variety of reef material makes for an interesting scuba diving and fishing location. Divers can swim through large culverts, examine what remains of the fuel tanks, and explore a small barge all in one dive. At least one resident Goliath Grouper always seems to be on site.

3) PG Barge

GPS Coordinates: *30° 20.320'*
                                                 81° 11.790'

Depth: *70 feet*

Structure: *Metal Barge*

Distance and Heading
Jacksonville Inlet: *11.55 miles @ 110.97°*
St Augustine Inlet: *29.87 miles @ 009.80°*

Description:

The PG Barge Reef is close enough to the Jacksonville In-let to make for a partial day trip. The metal reef is home to many creatures that make it worth a visit. In fact, during our last visit, we encountered a massive Manta Ray with a wing span of what seemed to be 20 feet. The Manta circled over the barge throughout our entire dive, casting a giant shadow with every pass.

4) **Natural Ledge 2**

GPS Coordinates: *30° 20.628'*
                                                       81° 11.411'

Depth: *70 feet*

Structure: *3 foot ledge*

Distance and Heading
   Jacksonville Inlet: *11.78 miles @ 108.70°*
   St Augustine Inlet: *30.29 miles @ 010.39°*

Description:

This reef is a small natural ledge creating three to four feet of relief off the sea floor. The exposed ledge is heavily encrusted with beneficial organisms providing food and shelter for a broad range of marine animals. One interesting thing about this ledge and "Natural Ledge 1," is that there are small caverns underneath certain sections of the ledges. Marine life often encountered includes Black Sea Bass, Oyster Toadfish, ornamentals, Gag Grouper, sea turtles, and much more.

5) Natural Ledge 1

GPS Coordinates: *30° 20.981'*
81° 10.960'

Depth: *70 feet*

Structure: *3 foot ledge*

Distance and Heading
Jacksonville Inlet: *12.09 miles @ 106.19°*
St Augustine Inlet: *30.77 miles @ 011.07°*

Description:

The Natural Ledge 1 site is in close proximity to Natural Ledge 2. There are a few patches of live bottom and small ledges in this area, and each site has similar characteristics. The three foot ledge is heavily encrusted with invertebrate growth creating habitat for marine diversity. Like Natural Ledge 2, you are likely to find small caverns underneath ledge overhangs.

Page 23

6) Banana + Pogey Boats

GPS Coordinates: *30° 20.201'*
                                                   81° 11.183'

Depth: *70 feet*

Structure: *Metal vessels*

Distance and Heading
Jacksonville Inlet: *12.19 miles @ 110.50°*
St Augustine Inlet: *29.85 miles @ 011.04°*

Description:

This reef is actually two reefs in one. A banana boat sits perpendicular to a pogey boat; so close the vessels are nearly touching. Much of the banana boat has deteriorated away, but the two vessels together still suffice for productive marine habitat. Only 12 miles from the Jacksonville Inlet, this is a great spot for a partial day fishing or diving trip.

The top picture to the right shows reef building pioneer Ed Kalakauskis near the screw of the banana boat.

7) Gulf America Wreck

GPS Coordinates: *30° 16.670'*
                                         81° 13.630'

Depth: *60 feet*

Structure: *WWII Shipwreck*

Distance and Heading
    Jacksonville Inlet: *12.22 miles @ 132.97°*
    St Augustine Inlet: *25.57 miles @ 007.53°*

Description:

The SS Gulf America tanker was positioned roughly 9 miles east of Jacksonville Beach on the evening of April 10th, 1942, when she was hit by a torpedo from a German U-boat. The tanker sank and lives were lost. The story of the SS Gulf America is still being written, and today she represents the good that can come from the bad. What remains of the tanker is now a thriving reef with coral heads far larger than those found on other reefs off our coast. It is an amazing fishing and scuba diving location. Marine life seems endless and in great diversity. Be sure to visit this thriving habitat and historical site.

Jacksonville Reef

8) Tugboat #1

GPS Coordinates: *30° 19.810'*
                                        81° 10.980'

Depth: *70 feet*

Structure: *Metal Tugboat*

Distance and Heading
    Jacksonville Inlet: *12.51 miles @ 112.16°*
    St Augustine Inlet: *29.46 miles @ 011.54°*

Description:

This reef is a metal tugboat sitting in approximately 70 feet of water. The area is home to various small fish species including Tomtate Grunt, White Spotted Soapfish, Belted Sandfish, and blennies. Larger fish are also found including Cobia, Vermilion Snapper, and Black Sea Bass. Some of the upper portions of the tug have deteriorated significantly, but the majority of the structure is largely intact providing significant relief off the seafloor.

9) School Bus Barge

GPS Coordinates:

30° 25.858'
81° 09.066'

Depth:

65 feet

Structure:

Barge and School Bus

Distance and Heading
Jacksonville Inlet: 13.63 miles @ 080.22°
St Augustine Inlet: 36.58 miles @ 012.18°

Description:

In the mid 90's, a barge was sunk with a school bus to create marine habitat in an area known as Busey's Bonanza (BB). The bus was strapped to the top side of the barge when she was deployed, but what is left of it now sits in the sand off to the side (of the barge). Today the barge is heavily encrusted with colorful corals and sponges. Fish of all kinds utilize the habitat here. Gag Grouper and Black Sea Bass are encountered and we have even spotted a sunfish (Mola Mola) during a winter dive. Proximity to the inlet makes this a reasonable reef to visit.

10) Gibbs Dry Dock

GPS Coordinates: *30° 25.875'*
                                                         81° 08.310'

Depth: *75 feet*

Structure: *Wooden Dry dock*

Distance and Heading
  Jacksonville Inlet: *14.35 miles @ 080.92°*
  St Augustine Inlet: *36.52 miles @ 013.18°*

Description:

This reef is what remains of an old wooden dry dock that was placed decades ago. Wooden structure has disappeared leaving behind vertical spikes of metal. The unrecognizable ruble acts as a productive reef system.   Large Cobia are reported to be sighted by scuba divers and fishermen regularly during the summer months.

11) Press Box Ledge Reef

GPS Coordinates: *30° 23.856'*
                                    81° 03.930'

Depth: *80 feet*

Structure: *Natural Ledge Reef*

Distance and Heading
   Jacksonville Inlet: *18.61 miles @ 090.12°*
   St Augustine Inlet: *35.92 miles @ 021.00°*

Description:

This reef consists of a natural ledge system very close to the Press Box site. The ledge is roughly three to four feet in height off the sea floor and it is covered in marine life. As the pictures to the right show in detail, every square inch of this ledge is encrusted with beneficial growth. Some of the typical fish species encountered include Black Sea Bass, Beaugregory Damsels, Blue Angelfish, Sheepshead, and Nurse Sharks. This is a great reef to visit for fishermen, scuba divers, and marine life observers.

Page 35

12) Press Box

GPS Coordinates: 30° 23.823'
                                                                81° 03.578'

Depth: 80 feet

Structure: Metal Frame

Distance and Heading
Jacksonville Inlet: 19.20 miles @ 090.17°
St Augustine Inlet: 36.12 miles @ 021.89°

Description:

In the early 1980's several sections of metal press boxes, from the former Jacksonville Gator Bowl, were deployed offshore to create an artificial reef. Today the structures continue to function as a productive marine habitat. Much of what remains includes metal framing encrusted with coral, sponge, and algae. The usual fish suspects include Tomtate Grunts, Black Sea Bass, Grey Triggerfish, Blue Angelfish, and the occasional Cobia. This is a well known and popular fishing and diving reef system.

13) CH Tug

GPS Coordinates: *30° 18.560'*
                                            81° 04.150'

Depth: *80 feet*

Structure: *Tugboat*

Distance and Heading
Jacksonville Inlet: *19.39 miles @ 108.48°*
St Augustine Inlet: *30.23 miles @ 024.77°*

Description:

This reef is composed of a sunken metal tugboat sitting near a natural ledge reef. The tug has scoured its way into the seafloor over time, providing a great dug-out area near the propeller. Large Gag Grouper tend to congregate in the hole created.

14) FF Concrete 2011

GPS Coordinates: *30° 10.296'*
                                        81° 09.069'

Depth: *75 feet*

Structure: *Concrete Culverts*

Distance and Heading
    Jacksonville Inlet: *20.66 miles @ 139.25°*
    St Augustine Inlet: *19.56 miles @ 023.49°*

Description:

In July of 2011, nearly 700 tons of surplus and recycled concrete was placed at this reef site. The material includes large culvert pipes and box junction type structures. Even though it is a relatively new reef, the area is already an oasis of marine life. The concrete is beginning to encrust with growth and several species have already been observed. Both this site and the nearby "FF Concrete 2009" reef are known to hold Summer Flounder and Cobia in the summer months, as well as the typical north Florida reef inhabitants.

Jacksonville Reef

15) FF Concrete 2009

GPS Coordinates: *30° 10.034'*
                                        81° 09.322'

Depth: *75 feet*

Structure: *Concrete Culverts*

Distance and Heading
   Jacksonville Inlet: *20.73 miles @ 140.33°*
   St Augustine Inlet: *19.18 miles @ 023.16°*

Description:

Nearly two years prior to the deployment of the FF Concrete 2011 Reef, another reef (FF Concrete 2009) was created. Like the 2011 project, FF Concrete 2009 was created using between 700 and 800 tons of recycled concrete pieces. Most of this material includes giant box junctions and culvert pipes. The concrete is already heavily encrusted with growth, allowing for a dense and diverse population of fish species. Divers report frequent sightings of Goliath Grouper here.

16) HG Ledge

GPS Coordinates: 30° 13.012'
                                         81° 00.323'

Depth: 100 feet

Structure: 5 foot ledge

Distance and Heading
Jacksonville Inlet: 25.42 miles @ 119.32°
St Augustine Inlet: 26.72 miles @ 037.96°

Description:

This section of ledge rises approximately five to eight feet off the sea floor, and is located at an area fisherman call Hospital Grounds. There are several ledges and live bottom patches near this particular section of ledge. The ledge holds a variety of marine life, some of which is shown in the pictures to the right. Other creatures often encountered include Gag Grouper, Red Snapper, and large Spiny Lobster.

17) Coppedge Tug

GPS Coordinates: 30° 17.361'
                                          80° 57.887'

Depth: 80 feet

Structure: Metal Tugboat

Distance and Heading
  Jacksonville Inlet: 25.76 miles @ 106.96°
  St Augustine Inlet: 32.22 miles @ 035.90°

Description:

In the summer of 1988, the 100 foot long Coppedge Tugboat was placed on the sea floor. Today she is serving as a thriving marine ecosystem, home to a countless amount of marine life. Most of the vessel is largely intact and completely encrusted with corals, sponges, algae, tunicates, barnacles, and much more. Vermilion Snapper, Mangrove Snapper, and Greater Amberjack are some of the more plentiful species targeted by fishermen. Very close to the tug reef is another artificial reef often referred to as the Coppedge Culvert Reef.

18) Coppedge Culverts

GPS Coordinates: *30° 17.290'*
                                                 80° 57.840'

Depth: *80 feet*

Structure: *Concrete Pieces*

Distance and Heading
    Jacksonville Inlet: *25.83 miles @ 107.10°*
    St Augustine Inlet: *32.18 miles @ 036.06°*

Description:

A great quantity of concrete material including culvert pipes, slab pieces, box junctions, and miscellaneous structures, was placed near the Coppedge Tugboat Reef in 1989. It is a fairly large reef area worth a visit for any scuba diver or fisherman. Some of the usual suspects include Mangrove Snapper, Carolina Hake, Gag Grouper, Blue Angelfish, and Grey Triggerfish.

19) The Spike

GPS Coordinates: 30° 22.529'
                                                  80° 53.689'

Depth: 110 feet

Structure: *Tender Tugboat*

Distance and Heading
    Jacksonville Inlet: *28.85 miles @ 093.04°*
    St Augustine Inlet: *39.49 miles @ 035.72°*

Description:

The Spike is a 70 foot long USCG Tender that was deployed in July of 2009. The vessel is nearly 35 feet tall providing significant relief and profile; perhaps attracting more pelagic fish than reefs with less profile. The interior walls and structure have been removed creating vast open spaces inside the vessel. Rumors exist of a large Goliath Grouper residing in the engine room. Common fish include Gag Grouper, Spadefish, Greater Amberjack, Great Barracuda, and Tomtate Grunt. Fishermen have reported the occasional White Shark in this area over the winter months.

20) Bob Engel

GPS Coordinates: *30° 21.630'*
                                        80° 53.240'

Depth: *110 feet*

Structure: *Massive Barge*

Distance and Heading
    Jacksonville Inlet: *29.37 miles @ 095.02°*
    St Augustine Inlet: *38.93 miles @ 037.14°*

Description:

The nearly 300 foot barge that makes up the Bob Engel Reef was deployed in 2001. Although she sits in over a hundred feet of water, the vessel provides ten to fifteen feet of relief off the sea floor. There is plenty of sea life to be found here including Tomtate Grunts, Gag Grouper, Greater Amberjack, and various snapper species. This is a great location for the advanced diver or the offshore bottom fisherman.

21) BR Ferry

GPS Coordinates: *30° 21.714'*
                                                      80° 49.975'

Depth: *110 feet*

Structure: *Metal Ferry*

Distance and Heading
    Jacksonville Inlet: *32.60 miles @ 094.32°*
    St Augustine Inlet: *41.06 miles @ 040.65°*

Description:

Over 30 miles from the Jacksonville Inlet, this reef sits in deep enough water to limit scuba divers to those with advanced certifications and above. However, it is a worthwhile fishing and diving location to visit as it holds a great diversity of sea life. Reefs in this area have tended recently to include high populations of the invasive Lionfish. Divers and fishermen are encouraged to carefully harvest these detrimental animals from these reefs. It turns out their meat is tasty and marine conservation organizations are encouraging a demand for it to help with eradication efforts. See "Invasive Lionfish" on page 196

Page 55

22) Casablanca

GPS Coordinates: 30° 17.507'
                                        80° 49.318'

Depth: 115 feet

Structure: 330' Landing Ship

Distance and Heading
Jacksonville Inlet: 33.97 miles @ 102.42°
St Augustine Inlet: 38.02 miles @ 046.18°

Description:

The massive Casablanca is an over 300 foot vessel that was sunk for artificial reef creation in 1972. Ironically, the ship resides several miles south of where she was intended to sink. Sources say that weather was an issue during the deployment and the reefing crew departed too soon. The vessel floated just under the surface of the ocean unnoticed until she sank at her current location Today, the Casablanca is a large reef site that houses many animals. Unfortunately this reef, like other deeper water reef sites in the area, has become a popular site for invasive Lionfish.

ST. AUGUSTINE

St. Augustine
Inlet

St. Augustine Florida Reefs

Reef Map #	Reef Name	Page #
15	FF Concrete 2009	42
14	FF Concrete 2011	40
23	Andy King	62
24	9-Mile North Culvert	64
25	Standish Reef	66
26	High School Reef	68
27	Desco	70
28	Desco Shrimp Boat	72
29	Little Barge Culverts	74
30	Little Barge Reef	76
31	Dorthy Louise Barge	78
32	Mantanzas Barge	80
33	Grady Prather	82
34	Moody Concrete	84
35	Taylor Reef	86
36	Reggie Tug	88
37	Intruder Aircraft	90
38	Capo	92

St. Augustine
Inlet

14
15
26
25
24
23

10 miles

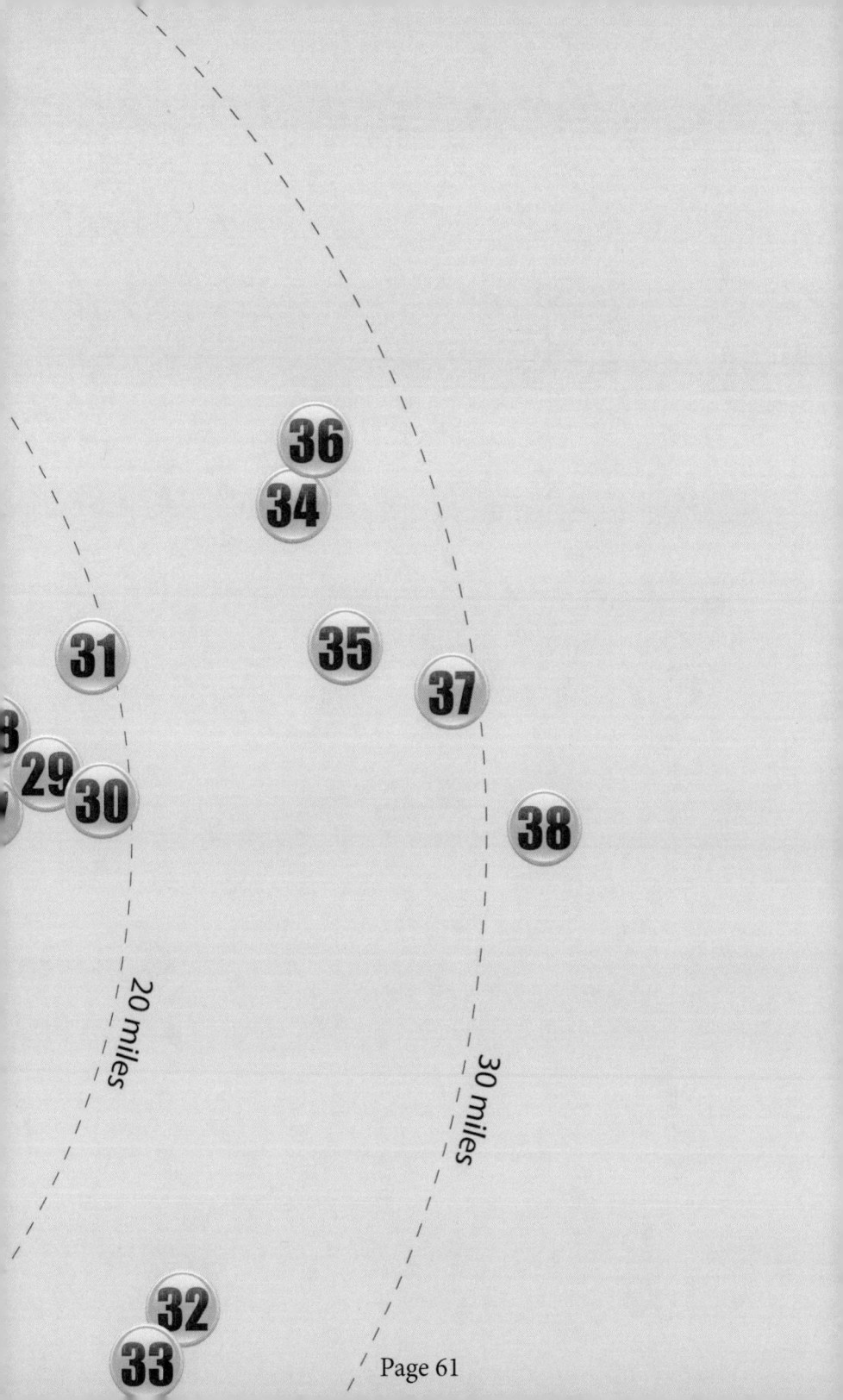

36

34

31

35

37

8

29

30

38

20 miles

30 miles

32

33

23) Andy King

GPS Coordinates: *29° 52.591'*
                                 81° 09.207'

Depth: *60 feet*

Structure: *Concrete Pilings*

Distance and Heading
   St Augustine Inlet: *8.08 miles @ 107.38°*
   Jacksonville Inlet: *38.37 miles @ 159.67°*

Description:

While the Bridge of Lions in St. Augustine was undergoing extensive repairs, a temporary bridge was constructed. After the repairs were completed, the temporary bridge was dismantled and the pilings were placed here in July of 2010. Although a relatively new reef it is already home to Black Sea Bass, Grey Triggerfish, and bait fish. A short distance from the St. Augustine Inlet makes this an easy reef to access.

ANDY KING REEF
ESTABLISHED
30 JULY 2010
BY A.C.G.F.A.

24) 9-Mile North Culverts

GPS Coordinates: *29° 54.670'*
                                          81° 07.180'

Depth: *60 feet*

Structure: *Concrete Culverts*

Distance and Heading
    St Augustine Inlet: *9.74 miles @ 090.07°*
    Jacksonville Inlet: *36.94 miles @ 155.45°*

Description:

Less than 10 miles from the St. Augustine Inlet, the 9-Mile North Culvert Reef is easy to access. The reef was created in 1983 and has been maturing into productive habitat for decades. Due to its accessibility, this site is known to draw traffic on the weekends and holidays. Try to visit on an off day for the best experience. This is a geat place to catch fish close to the inlet.

25) Standish Reef

GPS Coordinates: *30° 01.247'*
                                             81° 05.847'

Depth: *70 feet*

Structure: *Concrete Boxes*

Distance and Heading
   St Augustine Inlet: *13.51 miles @ 055.84°*
   Jacksonville Inlet: *30.89 miles @ 147.43°*

Description:

Standish Reef is composed of concrete junction boxes and is within the fifteen mile trip mark from the St. Augustine Inlet. Plenty of Black Sea Bass and Grey Triggerfish can be found here. As the picture to the right demonstrates, the surface area of the concrete material serves as a hold-fast for spectacular soft corals to flourish. This type of coral would not exist in this location without a solid substrate like the concrete provides.

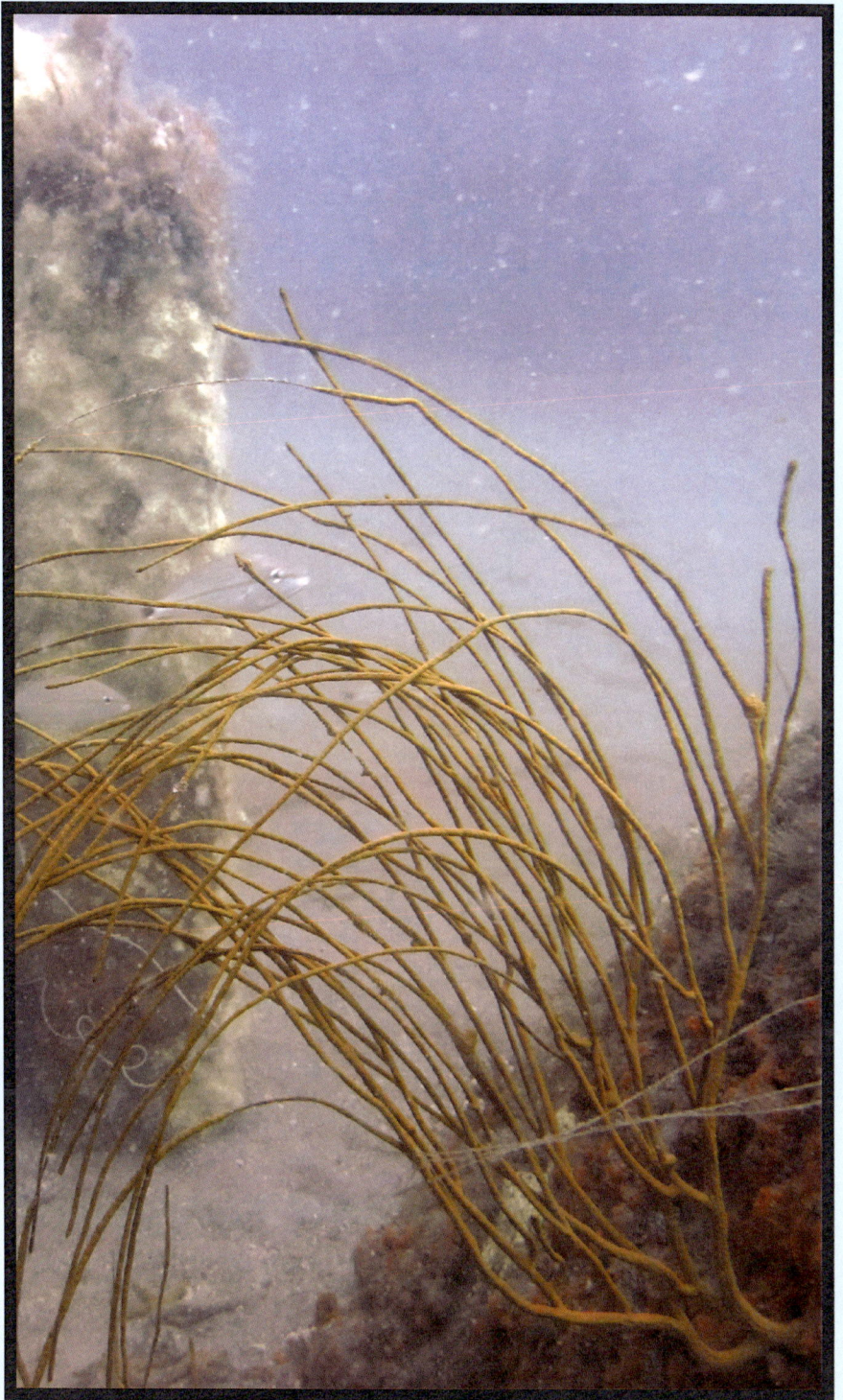

26) High School Reef

GPS Coordinates: *30° 03.559'*
                                        81° 06.450'

Depth: *70 feet*

Structure: *Reef Balls + Concrete*

Distance and Heading
St Augustine Inlet: *14.56 miles @ 045.83°*
Jacksonville Inlet: *28.35 miles @ 145.53°*

Description:

The High School Reef is an artificial reef site created from concrete structures. The photos to the right show images from the reef ball section of the reef. The structures are heavily encrusted with a variety of sponges. The yellow, pink, orange, and grey colors in the pictures are different types of sponges observed. Black Sea Bass and Tomtate Grunts are the usual suspects at this reef.

27) Desco

GPS Coordinates: *29° 53.242'*
                              81° 00.960'

Depth: *70 feet*

Structure: *Concrete Culverts*

Distance and Heading
    St Augustine Inlet: *16.05 miles @ 095.91°*
    Jacksonville Inlet: *41.31 miles @ 148.54°*

Description:

The reef site known as Desco is an area of scattered concrete pieces, including culvert pipes and box junctions. The structures are heavily encrusted, primarily with healthy stony coral colonies. Plenty of fish can be found here beyond the Blue Angelfish and Polka-dot Batfish shown to the right.

28) Desco Shrimp Boat

GPS Coordinates: *29° 54.702'*
                                      81° 00.160'

Depth: *75 feet*

Structure: *Shrimp Boat*

Distance and Heading
    St Augustine Inlet: *16.76 miles @ 089.88°*
    Jacksonville Inlet: *40.33 miles @ 146.33°*

Description:

The Desco Shrimp Boat Reef is what remains of a sunken shrimp boat. It is a relatively small reef site but holds plenty of sea life. Divers may be surprised to encounter metal hardware on the wreck that appears almost brand new. Many apparatuses on a shrimp boat include high grade stainless steel components. These pieces are often shiny with no corrosion even though they have been submerged for as long as the rest of the eroded vessel.

29) Little Barge Culverts

GPS Coordinates: *29° 53.620'*
                                        80° 58.740'

Depth: *80 feet*

Structure: *Concrete Culverts*

Distance and Heading
    St Augustine Inlet: *18.22 miles @ 093.81°*
    Jacksonville Inlet: *42.15 miles @ 145.66°*

Description:

The Little Barge Culverts Reef was created in 1992 with concrete materials such as culvert pipes and box fittings. The pictures to the right are actually screen shots taken from video I was recording of the reef. You can see the reef has a mature appearance with a great diversity of life. Concrete slabs have evolved into beautiful soft coral gardens.

30) Little Barge Reef

GPS Coordinates: *29° 53.559'*
                                                              80° 58.530'

Depth: *75 feet*

Structure: *Metal Barge*

Distance and Heading
   St Augustine Inlet: *18.43 miles @ 093.98°*
   Jacksonville Inlet: *42.33 miles @ 145.48°*

Description:

The Little Barge Reef is also sometimes referred to as the 20' X 60' Barge. Although a smaller barge, this reef is big on marine life. A few of the pictures to the right show just a hint of the diversity of life found here. From elaborate stony corals to ornamental fish, this little reef is swarming with life. Unfortunately this also includes invasive Lion-fish, but the reality is these invaders are now observed on practically all north Florida reefs.

31) Dorothy Louise Barge

GPS Coordinates:

29° 56.316'
80° 57.508'

Depth:

80 feet

Structure:

Metal Barge

Distance and Heading
St Augustine Inlet: *19.50 miles @ 084.40°*
Jacksonville Inlet: *40.38 miles @ 141.74°*

Description:

The Dorothy Louise Barge Reef, also referred to as the DL Barge, is a nearly 200 foot long barge that is broken at one end. The structure creates a large reef footprint, creating habitat for a great variety of marine life. Popular animals encountered include sea turtles, Nurse Sharks, Almaco Jack, Goliath Grouper, bait fish, and the list goes on. A great reef to visit for the photography diver or the hungry fisherman.

Page 79

32) Matanzas Barge

GPS Coordinates:
                                  29° 40.689'
                                  80° 58.023'

Depth:
                                  70 feet

Structure:
                                  Metal Barge

Distance and Heading
    St Augustine Inlet: *24.86 miles @ 130.00°*
    Jacksonville Inlet: *55.35 miles @ 153.71°*

Description:

This reef and the Grady Prather Reef are included in the Flagler County reef jurisdiction, but since they are accessible by the St. Augustine Inlet I am including them here. This barge has a mature reef appearance and is heavily encrusted with branching stony corals. On our last dive this barge was loaded with juvenile Black Sea Bass. You can see a few in the pictures to the right.

33) Grady Prather

GPS Coordinates: 29° 40.419'
                                    80° 58.330'

Depth: 70 feet

Structure: Concrete

Distance and Heading
St Augustine Inlet: 24.91 miles @ 131.44°
Jacksonville Inlet: 55.60 miles @ 154.08°

Description:

Several hundred tons of concrete slab and piling material was placed here in the summer of 2011. Unfortunately, we have not been able to dive the site since the deployment and thus I do not have underwater photos of the reef. Judging by the marine life found on nearby reefs, I imagine the Grady Prather Reef is on its way to becoming a healthy and productive reef ecosystem.

Pictures shown are of the actual reef deployment.

Page 83

34) Moody Concrete

GPS Coordinates: 29° 59.129'
                                              80° 51.521'

Depth: 100 feet

Structure: Concrete Culverts

Distance and Heading
    St Augustine Inlet: 25.89 miles @ 078.49°
    Jacksonville Inlet: 42.08 miles @ 132.59°

Description:

An artificial reef composed of concrete pieces, including culverts, that lie in over 100 feet of water. This is a mature reef with a great diversity of marine life. Both Red Snapper and Mangrove Snapper are frequently sighted by divers. The reef makes for a great dive site for the advanced diver.

The pictures provided demonstrate how concrete structure serves as an ideal substrate for invertebrate life forms, such as corals and sponges, to grow and thrive.

St. Augustine Reef

Page 85

35) Taylor Reef

GPS Coordinates: *29° 55.691'*
                                       80° 50.731'

Depth: *100 feet*

Structure: *Concrete Culverts*

Distance and Heading
   St Augustine Inlet: *26.21 miles @ 087.36°*
   Jacksonville Inlet: *45.39 miles @ 135.57°*

Description:

Taylor Reef is made up of several clusters of concrete culvert piles and pieces. It is a popular location for fishing but is also known as a great scuba diving destination. The typical marine animals associated with artificial reefs of north Florida can be expected to be encountered. This could include sea turtles, sharks, Cobia, Greater Amberjack, Summer Flounder, Gag Grouper, and much more.

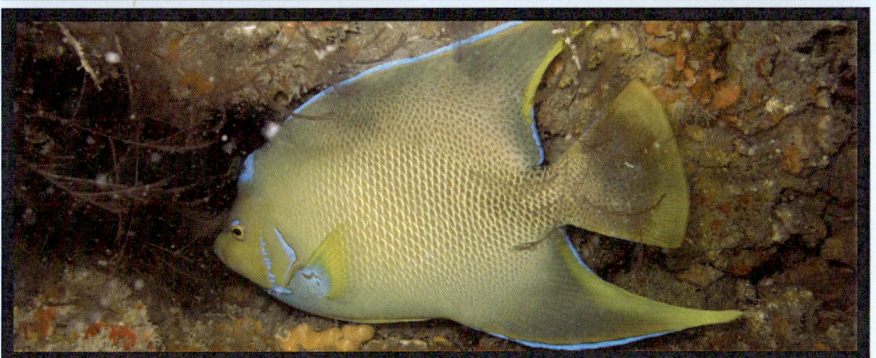

Page 87

36) Reggie Tug

GPS Coordinates: *30° 00.384'*
80° 50.873'

Depth: *100 feet*

Structure: *Metal Tugboat*

Distance and Heading
St Augustine Inlet: *26.87 miles @ 075.66°*
Jacksonville Inlet: *41.59 miles @ 130.46°*

Description:

The images to the right are screen shots from video footage I recorded of the Reggie Tug Reef on a poor visibility day. The imagery is not great by any means but demonstrates the marine habitat qualities of the reef. The remains of the tugboat are heavily encrusted with beneficial growth supporting many layers of the marine food chain.

St. Augustine Reef

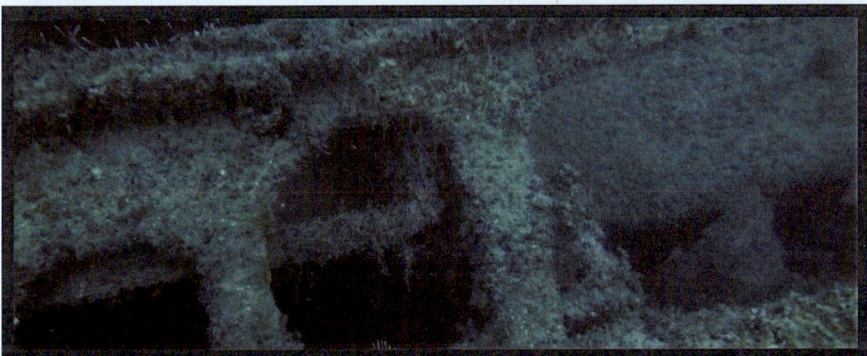

Page 89

37) Intruder Aircraft

GPS Coordinates: *29° 54.229'*
                                                    80° 47.996'

Depth: *100 feet*

Structure: *Aircraft Bodies*

Distance and Heading
St Augustine Inlet: *28.93 miles @ 090.93°*
Jacksonville Inlet: *48.51 miles @ 134.64°*

Description:

In June of 1995 over forty retired Navy A-6 Intruder air-craft bodies were placed off the coast of St. Augustine. The project was originated and coordinated by Steve Blalock of the St. John's County Reef Research Team. Blalock knew the clean metal Intruder frames would make great marine habitat, so he worked out the details with the Navy and the Department of Environmental Protection to make the reef project happen. Today this reef is a favorite among fishermen and scuba divers.

St. Augustine Reef

38) Capo

GPS Coordinates: 29° 51.046'
                                              80° 46.041'

Depth: 100 feet

Structure: Box Culverts

Distance and Heading
    St Augustine Inlet: 31.17 miles @ 097.62°
    Jacksonville Inlet: 52.48 miles @ 135.96°

Description:

Capo Reef is a deeper reef exceeding 100 feet in depth. The reef structures consist primarily of concrete box culvert sections. It is a great site for the advanced diver and the hook and line fisherman. Fish often encountered include Gag Grouper, Red Snapper, Great Barracuda, Lionfish, and a Goliath Grouper or two.

Page 93

MARINE LIFE

North Florida Marine Life

The offshore waters of northeast Florida include an enormous variety of life. It would take a whole encyclopedia collection of books to adequately describe such diversity. That amount of detail is outside the scope of this publication, but I have included the following chapter to shed some light regarding the more common creatures I have observed.

The following pages describe some regularly encountered specimens at north Florida reef sites. Unless otherwise mentioned, every photo has been captured at, or very near, a reef site offshore Jacksonville or St. Augustine. The descriptions provided are based on my experiences and observations with the marine animals in this region.

Almaco Jack

Seriola rivoliana

The Almaco Jack is often mistaken for the Greater Amberjack as both species look and act similar to each other. The differences are subtle, but the Almaco Jack is identifiable by a few features. Almaco Jack bodies are flatter and less elongated then the Greater Amberjack. Underwater the fish appears taller and stubbier then longer Greater Amberjack.

Almaco Jack are often encountered in schools in the water column around or above reef structure. In our diving experiences we observe these fish in the same places where we find Greater Amberjack. Most often seen in aggregations, occasionally a loan individual will be spotted swimming the reef.

Picture captured at: "31) Dorothy Louise Barge" (page 78)

Anemone
Multiple species

The sea anemone is one of the many organisms that bring reef structures to life. They are found attached to rock, shells, concrete, or metal. The reef provides the anemone with a structure where it can adhere and sustain itself. Meanwhile certain reef inhabitants utilize the anemones as food or shelter.

Anemones are predators, catching zooplankton and small fish that float or swim within reach of their tentacles. Many species also benefit from the waste by-products of internal photosynthesizing zooxanthellae algae cells.

We find anemones as isolated individuals or in clustered groups. Sometimes they are extended with their tentacles spread open, and other times they are retracted into a slimy ball shape. When contracted, an anemone looks similar to a chewed piece of gum stuck to the underside of a school desk. Many times anemones are mixed in with other encrusting organisms like sponges, as demonstrated in the picture to the left.

Picture captured at: "24) 9-Mile North Culverts" (page 64)

Atlantic Deer Cowry

Macrocypraea cervus

In my opinion, this mollusk produces the most beautiful shell of all shell producing animals in the north Florida area. The shell comes in a naturally shiny polished form. The Atlantic Deer Cowry is a type of sea snail that gets its name from its white dots, which are similar to those found on a young deer. These creatures search the reef for algae to consume.

We often find empty cowry shells and living Atlantic Deer Cowries while diving reefs off Jacksonville and St. Augustine. More often than not, when we see a shiny exposed shell it turns out to be empty. Cowries have predators, which include other mollusks that will leave only the cowry shell behind. When we encounter living Atlantic Deer Cowries usually only a small portion of the bright shell is exposed while the rest is covered by the living mantle. When the mantle completely covers the shell the animal takes on a fuzzy porcupine appearance. This is a dramatically different look then the attractive shell itself.

Top picture captured at: "31) Dorothy Louise Barge" (page 78)

Bottom picture captured at: "21) BR Ferry" (page 54)

Atlantic Octopus

Octopus vulgaris

Octopuses have been called the most intelligent invertebrates in the ocean, and I think that is an understatement. In the lab setting these creatures have proved their ability to learn, showing they could figure out how to perform tasks such as opening up jars. Their physical abilities are synonymous with their level of intelligence. Octopuses can dramatically change their color, shape, skin texture, and movement style all for the sake of mimicry. They can seamlessly blend into their surroundings, or disguise themselves as other animals in both appearance and behavior.

One of my most interesting encounters with this animal was while diving a natural ledge system offshore Jacksonville. I was swimming along and spotted a glass bottle wedged in a crevice within the ledge. As I looked closer at the bottle, I a noticed a small Oyster Toadfish and thought I had an awesome eye for seeing the well camouflaged critter. Shortly after this I was startled as I watched the rock cluster above the bottle come to life, change color and shape, and dart away. I had just witnessed an Atlantic Octopus change instantly from a reef ledge disguise into a free swimming Black Sea Bass replica. It was remarkable how fast it all happened and how much the octopus looked and moved like a Black Sea Bass.

I have only observed a handful of octopuses in all of my underwater hours offshore north Florida. This is probably because they blend in so well. For you divers out there that hope to get lucky enough to spot one, I can offer one tip. Keep an eye out for an unnaturally piled assortment of shells as this may be the entrance to an octopus lair.

Picture captured at: "St. Augustine plane wreck" (not featured)

Altantic Spadefish
Chaetodipterus faber

Every reef offshore Jacksonville and St. Augustine is associated with schooling fish species. Some of the usual fish schools expected to be encountered include jacks, Barracuda, grunts, Cigar Minnows, and Atlantic Spadefish. The schooling characteristics differ among the types of fish. For example, jacks school in the water column above the reef while grunts school in very close proximity to, and often within, the reef structure. Atlantic Spadefish are somewhat unique in their behavior as they school in the water column above, around, and within the reef. During very calm seas, Atlantic Spadefish can be seen at the surface over a reef. They actually do what we call "finning," where they break the surface of the water exposing the upper portion of their bodies.

If you were to ask me what is the most likely "water column" schooling fish a north Florida diver might encounter, I would say the Atlantic Spadefish. Although you are likely to observe others as well, the Atlantic Spadefish is one fish I can almost promise you will find. I do not remember a dive on a reef where we have not been joined by Atlantic Spadefish.

Atlantic Spadefish eat invertebrates from the reef directly but we have also observed them eating jellyfish. They average around twelve inches in length and are considered to be a good food fish.

Pictures captured at: "17) Coppedge Tug" (page 46)

Beaugregory Damsel
Stegastes leucostictus

North Florida reefs are actually home to a variety of colorful fish that are often associated with the Caribbean and more tropical waters. The Beaugregory Damsel is one example, and often the first vibrantly painted fish we encounter on our dives. They inhabit the many nooks and crannies of the reef, darting around hunting for any creature small enough to fit in their mouths. Algae grazing is also a part of their routine and diet.

During its younger state, when the fish is less than a couple inches in length, the damsel is a brightly colored blue and yellow. This is how we most often observe the Beaugregory Damsel. As the fish age, their bright color pattern tends to dull. Adults are about four inches in length and usually brownish in appearance.

Picture captured at: "10) Gibbs Dry Dock" (page 32)

Belted Sandfish

Serranus subligarius

This little fish is found on just about every reef we dive offshore Jacksonville and St. Augustine. Although averaging less than 2 inches in length they are in the same family (Serranidae) as grouper. They are bold predatory fish, eating critters smaller then themselves.

While diving we watch the behavior of the Belted Sandfish. They sporadically dart in and out of reef structure, pausing momentarily every so often as if to be making their own observations. The fish take great interest in anything we might stir up off the bottom when moving around. It is assumed they are hoping we unearth an easy meal for them.

We see Belted Sandfish more frequently on mature artificial reefs and natural ledge reefs. The fish are occasionally caught on rod and reel with use of a Sabiki Rig or small baited hook.

One observation of concern regarding the species is that it appears we are witnessing far fewer Belted Sandfish on reefs inhabited by lionfish.

Pictures captured at: "10) Gibbs Dry Dock" (page 32)

Black Sea Bass

Centropristis striata

I can confidently say that the Black Sea Bass is one of the most popular fish we encounter during our underwater expeditions offshore northeast Florida. We find them on just about any site that includes structure. Regardless if it is natural live bottom or an artificial reef, Black Sea Bass are present.

We not only encounter Black Sea Bass on practically every dive, but we also find them in great numbers. In fact, Black Sea Bass are often so thick in quantity that we are unable to observe the structure they are crowding. This is often frustrating because our underwater efforts require us to be able to photograph reef structures. We are forced to spend valuable underwater time trying to encourage the fish away from the structure we are to capture images of. It never works!

Black Sea Bass are very curious. They will come right up to our masks as if they are looking at themselves in the reflection. They also take great interest in our equipment such as cameras, tools, and dive gear. This curiosity sometimes evolves into aggression or hunger. The fish will begin to bite at anything. This includes movable parts on equipment as well as fingers, faces, and hair. During a recent dive I was watching a Black Sea Bass that just grabbed my hair, swim over to my dive partner Ed Kalakauskis and bite him square on the cheek! I have never laughed so hard underwater before!

See the picture of the buried Black Sea Bass. This is a behavior neither I nor my colleges have observed before.

Top picture captured at: "14) FF Concrete 2011" (page 40)
Middle picture captured at: "16) HG Ledge" (page 44)
Bottom picture captured at: "St. Augustine plane wreck" (not featured)

Blennies
Multiple species

Top Left: Seaweed Blenny, *Parablennius marmoreus*
Top Right: Saddled Blenny, *Malacoctenus triangulatus*
Bottom: Molly Miller Blenny, *Scartella cristata*

Blennies are some of the smallest fish inhabiting north Florida reefs but they may have the biggest personalities. Averaging one inch in length and less, they are sometimes hard to spot for a scuba diver. The inquisitive little fish will most definitely notice you before you observe it. The large eyes on its tiny face will follow your movements in a way that makes it appear the blenny is making facial expressions. Blennies are incredibly entertaining to watch. After they are bored with you they will go back to their normal blenny routine, darting around reef structure picking at bits here and there.

I have encountered blennies on every reef dive I have made in the area (excluding survey dives on freshly placed artificial reefs). They are everywhere there is structure with growth. Blennies are sporadic, darting around and then pausing momentarily to perch. Occasionally one will be encountered comfortably still in a sheltered crevice or hole, but for the most part they do not stay motionless long.

There are several types of blennies in the offshore waters of Jacksonville and St. Augustine. I have included pictures of three. The Seaweed Blenny and Molly Miller are often confused as they are similar in appearance at various stages in their life. Some blennies have different appearance phases. The yellow phase of the Seaweed Blenny shown is drastically different from its darker state.

Top pictures captured at: "30) Little Barge Reef" (page 76)
Bottom picture captured at: "17) Coppedge Tug" (page 46)

Blue Angelfish
Holacanthus bermudensis

The Blue Angelfish is arguably one of the most beautiful fish encountered in the western Atlantic region and it is no stranger to northeast Florida. We usually encounter at least one pair of Blue Angelfish during every dive, and yes, they are most often encountered in pairs. The average sizes observed are between ten and thirteen inches in length. Occasionally we find tiny juveniles, which have a different color pattern then the adults, hiding in crevices within reef structure.

Blue Angelfish swim freely but stay very close to reef structure. They graze primarily on the abundance and variety of encrusting sponges. Their magnificent colors make for a great photography specimen, and they will usually allow a diver to approach slowly.

Top picture (adult) captured at: "27) Desco" (page 70)
Bottom pictures (juveniles) captured at: "10) Gibbs Dry Dock" (page 32)

Carolina Hake

Urophycis earllii

Carolina Hake are truly strange looking creatures; I consider the species the "Centaur" of the ocean. It has a head section looking similar to that of a scale-less drum or trout, and an eel-like tail section. Where normal pelvic fins would protrude on a typical fish, instead appear leg-like structures. These modified pelvic fins (legs) allow the Carolina Hake to search for food in the sand.

We encounter Carolina Hake offshore, most frequently, at concrete culvert artificial reefs. They are not the easiest animals to find as they tend to take deep shelter inside cracks and ledges created by concrete pieces. Usually they are motionless and appear to be resting. We imagine the fish are more active in the overnight hours moving about looking for food.

The average size is in the eight to fourteen inches range. Both images provided to the left were captured at Jacksonville's Coppedge culvert reef.

Pictures captured at: "18) Coppedge Culverts" (page 48)

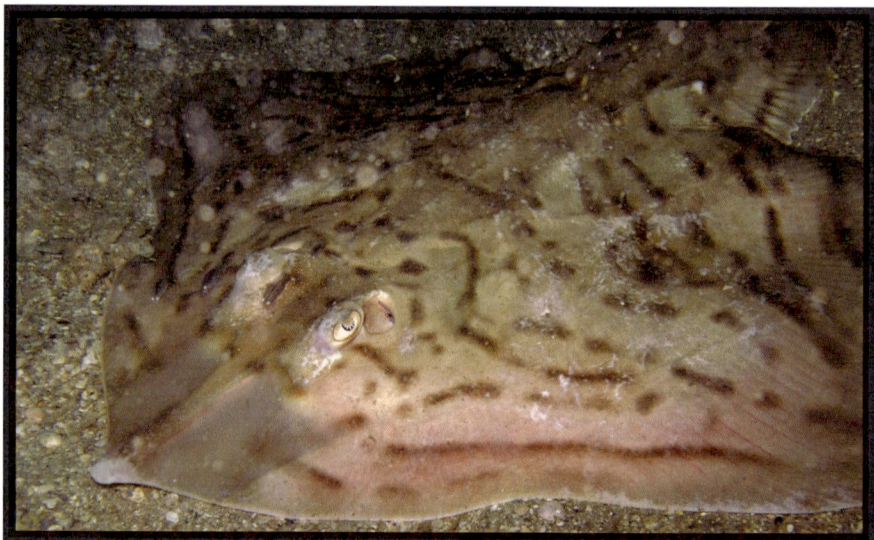

Clearnose Skate

Raja eglanteria

The Clearnose Skate must be named so because of the translucent section of flesh just in front of its eyes. Skates are often confused with stingrays and are mistakenly feared to have a venomous barb attached to their tail section. Skates, although somewhat similar in appearance, do not have such barbs. Some skates do however have pointed protrusions on their bodies. Although not venomous, the sharp edges could puncture human flesh if the animal was accidentally stepped on.

Clearnose Skates are found in the inshore and near shore waterways of north Florida. The two pictures of the skate provided to the left were captured at an artificial reef site twelve miles off the coast of Ponte Vedra during the winter. Skates eat shrimp, crabs, and small fish. They are responsible for the "mermaids purses" we often find offshore. The purses are actually the eggs laid by the skates. The middle picture to the left is one such egg that has appeared to have already hatched.

Pictures captured at: "14) FF Concrete 2011" (page 40)

Crabs
Multiple species

North Florida reefs are home to many types of crabs. These creatures utilize the reefs for food and shelter, while at the same time helping the reef. Crabs often serve as a food source to fish larger then themselves but also maintain the reef with their activities. Many crabs consume decaying organic matter, algae, and other inverts that could potentially overwhelm the reef if not kept in check. Below are two examples of local crabs we encounter while underwater.

Top image:

Bocourt Swimming Crab

Callinectes bocourti

This is a female crab with a cluster of eggs protruding from her abdomen.

Bottom image:

Arrow Crab

Stenorhynchus seticornis

A common crab on our reefs but often difficult to locate because of its camouflage and size. They are jokingly referred to as the "Daddy Long Legs" of the sea.

Crabs "Cont"
Multiple species

Top image:

Hermit Crab

Possibly: *Petrochirus diogenes*

      Hermit crabs are unique to typical crabs as they carry a shell home. They lack the tough rigid protection throughout their entire body like that of a normal crab, and make for an easy meal if not protected. The shell that is carried provides the exoskeleton shelter that is missing. A variety of hermit crabs are found throughout the local offshore waterways, grazing the reef structures for food. Hermit crabs have similar food habits as traditional crabs and are opportunistic. They will consume whatever food sources are available.

Bottom image:

Texas Longhorn

Genus species?

      The Texas Longhorn is a unique specimen. It is actually a colony of Bryozoa that houses a crab. The entire "shell" is a living mass with a crab that carries it around. The picture to the left is the only time I have ever encountered the Texas Longhorn and its mutualistic hermit crab partner. This crab and bryozoa package was encountered near the FF concrete reef sites.

Picture captured near: "14) FF Concrete 2011" (page 40)

Florida Horse Conch

Triplofusus giganteus

The Florida Horse Conch is the largest sea snail inhabiting offshore north Florida waters. Although the snail can reach greater lengths, we typically find them in the six through eight inches range. They are predatory and hunt the reefs in search of other sea snails to consume.

We encounter at least one Florida Horse Conch during just about every reef dive. The snail is found on reef structures or on the sandy sea floor. Look closely at the bottom picture towards the left. The Florida Horse Conch is actually on the back of a Summer Flounder.

Top picture captured at: "34) Moody Concrete" (page 84)
Bottom picture captured at:: "1) MR East Barge" (page 14)

Forbes Sea Star

Asterias forbesi

The Forbes is an abundant sea star offshore north Florida. We find several of the starfish on practically every dive investigation of a reef site. They slowly search the structure looking for bivalve mollusks to consume. When a scallop or an appropriate meal is found, the sea star is able to pry it open enough to insert its (the sea star's) stomach. Once the stomach is inserted, enzymes start to digest the prey inside its own shell.

Normally we encounter the Forbes independently with a five leg configuration. Occasionally however, we find clusters of the sea star as shown in the image to the left. If you look to the far left you will see a six legged specimen.

Pictures captured at: "5) Natural Ledge 1" (page 22)

Gag Grouper
Mycteroperca microlepis

Gag Grouper are one of the most sought after food fish species by both recreational and commercial fisherman in this region. Their meat is considered delicious and as a large fish each Gag Grouper provides pounds of flesh. The intense attraction to harvest grouper translates to heavy fishing pressures. This causes concern for the overall health of Gag Grouper populations, and so regulatory measures are in place to attempt to control the amount fish harvested. These rules tend to change frequently and should be reviewed prior to every offshore trip for those considering harvesting grouper.

We encounter a small number of Gag Grouper during most underwater reef explorations offshore north Florida. The fish observed range from as small as twelve inches in length to over three feet long. They are curious and intimidated by divers at the same time. Generally they will disappear as a diver approaches.

This past summer (2012) concerns were raised as numbers of Gag Grouper were observed in a sickly state. Divers reported finding grouper lying on their sides, acting near death, and with what appeared to be abrasions on their bodies. These incidents were reported after a period of tropical weather where the region received significant rainfall. Some fishermen have claimed to see similar behaviors historically coinciding with heavy rain producing weather. They argue that the lower salinity might be to blame. To date, I am unaware of a determined conclusion explaining the sickly grouper. We have not however received further reports of sick fish since then.

Top picture captured at: "13) CH Tug" (page 38)
Bottom picture captured at: "19) The Spike" (page 50)

Goliath Grouper
Epinephelus itajara

The name Goliath Grouper tells us all we need to know about this animal. It simply is a massive grouper. Average specimens weigh over several hundred pounds with some extreme cases as large as 800 pounds (according to the Coleman & Koenig Laboratory of Florida State University).

Goliaths are a relatively common site to divers in this area of Florida. Often one or two of these grouper will be encountered at larger artificial reef sites. Many times you can hear and feel them before you see them. Goliaths make this low pitch, deep boom, type sound that you can feel in your lungs when close enough. These fish look similar to other groupers sharing its range but are different in a few obvious ways. The head is much broader and more boxlike then a Gag Grouper, and a Goliath's tail fin is rounded. A Gag Grouper and other commonly observed groupers will have a flat tail fin. Goliaths of course mature at much larger sizes then other grouper species as well.

Pictures captured at: "Lee County Reefs" (not featured)

Great Barracuda
Sphyraena barracuda

During flat calm weather days, reef sites can literally be located by sighting barracuda just below the sea surface. These fish congregate over areas of structure to take advantage of the food source the reef below provides.

The Great Barracuda we generally encounter range from two to four feet in length and are found throughout the water column. More commonly, these torpedo shaped predators are observed hovering in aggregations near the sea surface, but we do find them near the reef structure hunting as well.

Great Barracuda have an opportunistic type diet, implying they eat what they can catch. I have witnessed these fish attack Bonita, Spiny Lobster, and even Summer Flounder. Fishermen complain of significant catch loss on reefs with heavy Great Barracuda populations. The opportunist fish are not shy about taking what fishermen are trying to reel in.

Great Barracuda are large, menacing looking predators, with very sharp teeth they proudly show off to divers. Having said that, we have never had an issue with these animals and we enjoy watching them during our safety stops. I would recommend divers avoid wearing anything flashy, such as jewelry, just to be on safe side. Great Barracuda eat shiny fish for a living so there is no need for divers to test their food recognition abilities. On more then one occasion, I have witnessed a large Great Barracuda bite completely through a mature Bonita like it was as soft as warm butter.

Pictures captured at: "31) Dorothy Louise Barge" (page 78)

Greater Amberjack
Seriola dumerili

The Greater Amberjack is one of the larger fish species associated with reefs offshore Jacksonville and St. Augustine. They are typically found in the water column above and around reef structure, and average in the twenty-four through thirty-six inch length range. Fishermen target the species for their strenuous fighting ability and for the quantity of meat they provide. Greater Amberjack is a local preferred seafood source to serve at fish cookouts.

From our diving experience, we expect to find Greater Amberjack at reef sites seventy feet and deeper. Reefs at these depths with higher profiles, especially metal vessels, seem to hold large schools of these fish at greatest frequencies. An ideal spot to encounter this jack is at the Spike Reef. The vessel sits at 110 feet deep but at over 35 feet in height it rises to a depth of 75 feet. This extreme vertical profile seems to be a magnet for the Greater Amberjack.

Pictures captured at: "19) The Spike" (page 50)

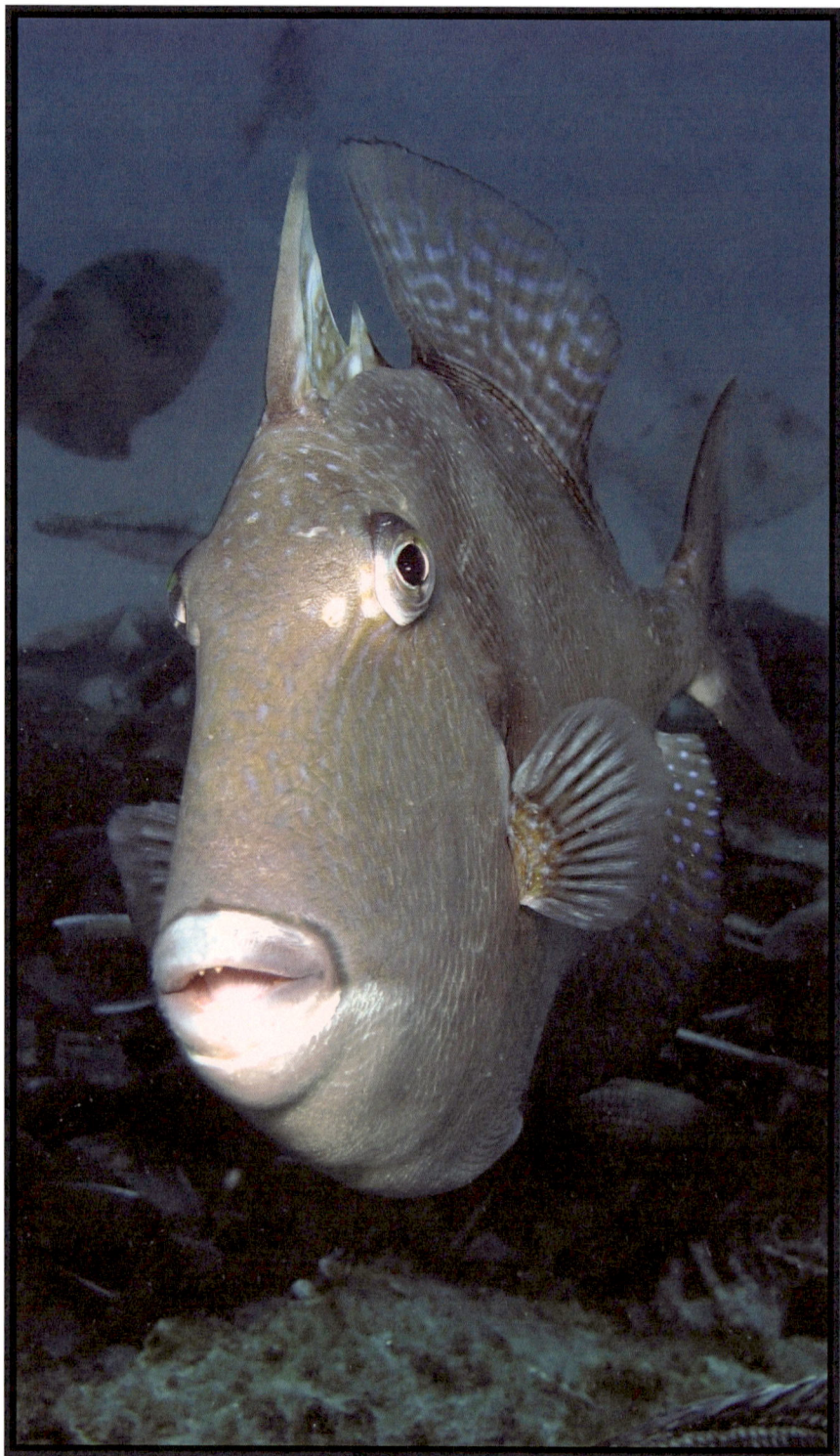

Grey Triggerfish
Balistes capriscus

The Grey Triggerfish may not be as colorful as some of the triggerfish found in the Pacific, but it is a tasty fish for those who enjoy seafood. Most anglers are familiar with these fish by having their bait stolen from the ends of their fishing hooks. Grey Triggerfish enjoy meaty foods offered by fisherman baits but have small hard mouths. This makes them difficult to catch by rod and reel. To further frustrate fisherman, Grey Triggerfish are found at the same offshore reef locations where other fish (such as Black Sea Bass and Gag Grouper) are targeted. Many times Grey Triggerfish will pick the bait clean off the hook before the targeted species has a chance to find it.

Normally we encounter less than twenty Grey Triggerfish per dive but on one occasion, during a dive at the freshly deployed Andy King Reef, we were surrounded by hundreds of them. Dive partner Ed Kalakauskis and I were conducting a survey of the reef and were amazed at the unexplainable numbers of these fish. Shortly into the dive I was watching Grey Triggerfish swarm around Ed when suddenly one, and then another one, bit him. Of course I found this quite entertaining and started laughing until I was bit on the ear. OUCH!! Maybe now is a good time to mention that Grey Triggerfish are equipped with mouths capable of crushing shell. So as you can imagine, their bites on human flesh are a bit more dramatic then a mosquito prick. We continued to get nipped and had to cut our dive short because we were having difficulties performing our tasks. This was a unique and unforgettable dive event, as we dive with Grey Triggerfish on most reefs without being attacked in this manner. Our guess is we stumbled across a breeding congregation and some of the fish were taking out their aggression on us.

Picture captured at: "St. Augustine plane wreck" (not featured)

High Hat
Equetus acuminatus

The High Hat is a type of drum similar in appearance to the Jackknife Fish which is also described in this book. Like most drum, it is a carnivorous fish seeming to prefer small inverts like crabs, shrimp, and worms. It is a common fish found on reefs throughout the western Atlantic and is also a species of interest to marine aquarists.

We encounter High Hats more frequently than the Jackknife Fish, spotting them on nearly every dive. Regardless of the reef structure we tend to find High Hats. Usually we encounter small aggregations in the four to six inch range, or as juveniles less than two inches in length. The silver and black color contrast is more dramatic on the juveniles as the larger specimens tend to darken in color.

Notice in the pictures provided how the stripe lines continue through the eyes. The High Hat is a smaller fish to find but a striking one to observe.

Pictures captured at: "5) Natural Ledge 1" (page 22)

Jackknife Fish
Equetus lanceolatus

As you might have guessed by looking at the picture of this fish, its common name might have something to do with its body shape. The dorsal and caudal fins are extremely pronounced creating the appearance of sharp edges. This unique shape and contrasting color pattern make this a sought after fish in the marine aquarium market.

If the fish in the image looks a little bit like a small drum, that is because it is in the same family as Red Drum, Black Drum, and other drum species. We find the Jackknife Fish on most reef dives. They are often in small schools swimming close to reef structure. Ironically, I think we encountered the largest schooling congregation of these fish at an airplane wreck we were investigating. The small wreck site consisted of two aircraft engines and a school of at least twenty Jackknife Fish hovering around. Even though the engines were partially buried, the fish stuck near them like they were magnets.

Picture captured at: "34) Moody Concrete" (page 84)

Moon Jellyfish

Aurelia aurita

Moon Jellyfish are found throughout most of the world's oceans including the offshore waters of northeast Florida. It is one of several varieties of jellies commonly encountered and is typically the largest. We find jellyfish frequently during our safety stops about twenty feet below the sea surface. As we wait several minutes off-gassing nitrogen, jellies will drift by in the current. During a dive we may see only one, but at other times the total amount of jellyfish is plentiful enough that we have to literary swim through them.

The jellies do not appear to congregate over particular reefs but pass over them incidentally. The prevailing currents bring them by. Jellyfish have pulsating mobility allowing them to move, but their overall navigation is largely current dependent. Occasionally we find Moon Jellies with tiny fish huddled closely inside them. The bottom picture on the left shows the juvenile fish taking refuge within the Moon Jellyfish.

Pictures captured above: "28) Desco Shrimp Boat" (page 72)

Moray Eels
Multiple Species

Morays are predatory animals consuming fish and invertebrates. Every now and then offshore bottom fishermen may catch one on line and reel, but it is much easier to capture a glimpse of an eel for a scuba diver. Some eels are very shy avoiding visual contact all together, while others will be curious with an approaching diver. It is advised to respect the space of a moray eel as their kind come equipped with menacing teeth and powerful jaws. We have never had any personal issues with these creatures but have heard of accounts where divers have been bitten. Most of these accounts however are the result of the diver acting inappropriately.

Spotted Moray Eel

Gymnothorax moringa

Spotted Morays are found on most reefs in the region but we tend to encounter the eels more frequently on metal vessel reef structures. The top and middle images to the left were captured at sunken ship artificial reefs. I am not sure there is a greater abundance of eels at the wrecks as they just may be more difficult to observe at other habitats with different structures. The top two images to the right are of Spotted Morays while the bottom picture is of another moray species.

Top picture captured at: "22) Casablanca" (page 56)
Middle picture captured at: "17) Coppedge Tug" (page 46)
Bottom picture captured at: "16) HG Ledge" (page 44)

Murex Snail

Hexaplex fulvescens

The Murex is another large sea snail found in the northeast Florida region. Although it does not get as large as the Florida Horse Conch it has similar habits. It preys on other mollusks often found at reef sites.

We tend to find these pointy snails in crevices and secluded places within reef structure. The size often observed is in the two to four inches length.

Picture captured at: "3) PG Barge" (page 18)

Nurse Shark

Ginglymostoma cirratum

The Nurse Shark is by far the most common shark we encounter while diving reefs offshore Jacksonville and St. Augustine. I would estimate we find one Nurse Shark on every reef dive. We usually encounter them resting motionless on the seafloor taking shelter underneath structure. Although this species can reach over ten feet in length, the sizes I have observed locally are generally less than five feet long.

Unlike pelagic sharks, the Nurse Shark can remain stationary and continue respiration. They are active during the night and rest during the day. This explains why we often find them motionless during our daytime explorations. Considering they are most likely asleep in this state, it is often possible to get very close to Nurse Sharks during daytime diving. Get too close however and a Nurse Shark may get annoyed and swim away. Nurse Sharks are considered a mild temperament shark but still need to be respected. Even though it is highly unlikely for this type of shark to bite a diver, their large size alone could be a hazard. Larger Nurse Sharks reach over ten feet in length and weigh over seven hundred pounds.

Pictures captured at: "35) Taylor Reef" (page 86)

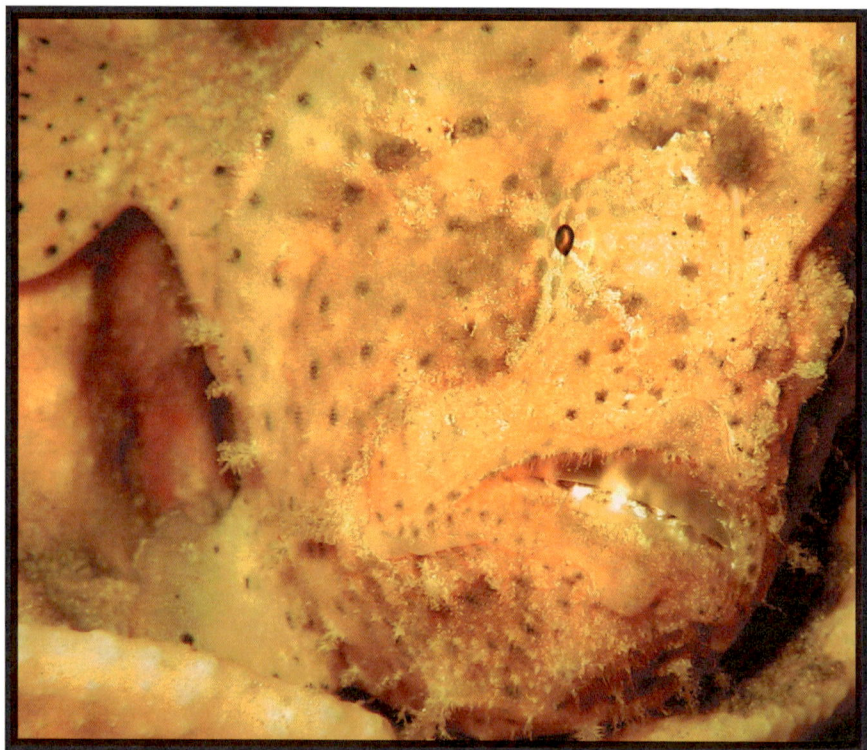

Ocellated Frogfish

Fowlerichthys ocellatus

Appropriately referred to as anglerfish, Ocellated Frogfish come equipped with built-in fishing poles and bait. If you look closely at the top picture of the yellow colored frogfish you will notice a small rod between its eyes capped off with a fuzz-like wad. This is the anglerfish's fishing rod and lure. When the fish is hunting, this little rod and lure will be set into action dancing about above its mouth. Eventually a curious prey fish will be drawn to the magic spell of the dance, and when it gets close enough it will be sucked in by a voracious gulp.

Frogfish are common in the western Atlantic including offshore north Florida waters. Having said that, I have only encountered a handful of frogfish sightings in my entire diving career. Frogfish are incredibly camouflaged blending in wonderfully efficiently with their surroundings. Furthermore, they lie motionless most of the time making them incredibly difficult to spot.

Top pictures captured at: "1) MR East Barge" (page 14)
Bottom picture captured at: "18) Coppedge Culverts" (page 48)

Oyster Toadfish

Opsanus tau

The Oyster Toadfish is often referred to as one of the uglier fish encountered on north Florida reefs. Understandably, it may not be the most attractive considering it is covered in mucus and is often speckled with wart-like growths. These traits however make this animal quite successful with the ability to live in a variety of water environments. In the Jacksonville area Oyster Toadfish are found offshore, in the intracoastal waterway , and in the St. Johns River.

My experience with these slimy beauties is at offshore reef locations like the ones mentioned throughout this book. They are very common although divers rarely see them because they blend into the surroundings so well. The "ugliness" of the toadfish is what allows it to disappear from its prey. Their colors and body protrusions make the Oyster Toadfish look like the surrounding reef structure and encrusting growth. A small fish only needs to mistakenly pass near the hidden Oyster Toadfish to become an entrée. Oyster Toadfish are ambush predators that wait in stealth mode until their prey swims by.

I often encounter Oyster Toadfish by mistake when attempting to photograph something else. Eventually I will unexpectedly get too close to the fish causing it to move in agitation. This is how I spot them; when they move. Other times I have grabbed a fish by mistake thinking it was part of the reef structure. As soon as you touch one you know what it is. Yuck!

Top picture captured at: "1) MR East Barge" (page 14)
Bottom picture captured at: "2) Culverts and Barge" (page 16)

Peacock Fire Worm

Possibly: Chloeia species?

One of the most exciting things I find about exploring reefs is the feeling of discovery when you encounter something you are not immediately familiar with. Even after all my years of scuba diving this continues to happen on each dive. Many times this is a unique structure, an animal behavior, or a new creature I have not seen before. Most of these personal discoveries are intriguing and have me wanting to get as close to the new object as possible to record observations. Occasionally there is an exception, and although I find most creatures of the planet ascetically pleasing, evolution has created some truly "icky" critters. Every now and then I find something on the reef fitting this category.

The first thing I thought about when I came across this worm (photo to left) was if my wet suit had any openings this thing could fit through. I actually found myself checking the fit of the sleeves around my wrist and the collar section around my neck. The last thing I wanted was a fuzzy worm crawling around stuck between my skin and thermal wear. The little hairy critter was speedily creeping along the seafloor with its thin flat body conforming to the shape of any object it passed over. It did not seem to have any interest in me or getting under my apparel.

This annelid (ringed worm) is possibly what is referred to in the marine aquarium industry as a Peacock Fire Worm. I included it in this publication solely to provide a sample of some of the extreme diversity encountered offshore north Florida.

Picture captured near: "15) FF Concrete 2009" (page 42)

Planehead Filefish

Stephanolepis hispidus

This filefish has a widespread range throughout the entire Atlantic Ocean. We occasionally encounter a solitary specimen investigating a reef structure or groups of smaller individuals hidden within floating sargassum clusters.

Sargassum is a type of floating macroalgae often found in small or very large patches drifting offshore. The drifting mats are floating reefs providing complex habit supporting many species. The Planehead Filefish is only one of the many critters that can be encountered. Many juvenile fish species, and even baby sea turtles will take refuge in the shelter provided by the sargassum.

Picture captured at: "19) The Spike" (page 50)

Polka-dot Batfish

Ogcocephalus cubifrons

This is truly one odd looking and behaving animal frequently observed offshore North Florida. From my personal diving experience I usually encounter at least one (or a pair) of these per reef visited. I usually find them sitting on the sea floor near reef structure. Often when I encounter a pair, they are seated face to face looking directly at each other. They just sit there with their faces only a few inches apart and stare. Very strange!

They do not appear to be bothered by the presence of scuba divers or camera equipment; lying completely still as I take my pictures only inches away. Every now and then one will attempt to swim away, but the movement is more similar to an assisted walk then a swim motion. They slowly skirt across the sea floor using their tail for propulsion and their modified pectoral fins like feet.

Most specimens I have encountered are less than 12 inches in length. Their body shape is somewhat bat-like with a triangular front section and rear tail section. Polka-dot Batfish mouths are at the very front of their head pointing downwards to aid in their hunting of small invertebrates.

Top picture captured at: "27) Desco" on page 70
Bottom picture captured at: "34) Moody Concrete" (page 84)

Porkfish

Anisotremus virginicus

━━━━━━━━━━━━━━━━━━━━━━━━━━━━━━━━━━━━━

A very popular grunt in south Florida and the Caribbean, Porkfish are also found offshore north Florida. The frequency of sighting these fish in this region is far less based on our diving experience. Further south it is typical to find large schools of Porkfish while our local encounters tend to be of lone specimens or isolated pairs.

The pictures featured to the left were captured at two different natural ledge reefs. The top picture was taken at HG Ledge while the bottom picture was captured on a ledge reef northeast of the Jacksonville Inlet. The latter ledge reef is not featured in this book but is in an area known as Rabbits Lair. It is a very high relief ledge that goes from about sixty to seventy feet in depth.

Top picture captured at: "16) HG Ledge" (page 44)

Red Snapper
Lutjanus campechanus

The Red Snapper, at the moment, is the most talked about fish in north Florida. It is a prized fish to recreational and commercial fisherman, but at the writing of this book is restricted to harvest. Scientists and regulatory agencies have concerns about the overall population of snapper and have set regulations in place to restrict harvest of the fish in this region. The majority of user groups that historically utilized the Red Snapper fishery disagree with the regulatory methods of population estimation and the resulting regulations put in place.

We generally encounter Red Snapper on the majority of reefs we dive. Typically we observe the Red Snapper in moderate size schools of twenty to forty fish, and do so most often at concrete structure artificial reefs. Sometimes we encounter smaller aggregations mixed in with groups of other fish, including Mangrove Snapper and Tomtate Grunts (as the picture to the right demonstrates).

Top picture captured at: "34) Moody Concrete" (page 84)
Bottom picture captured at: "18) Coppedge Culverts" (page 48)

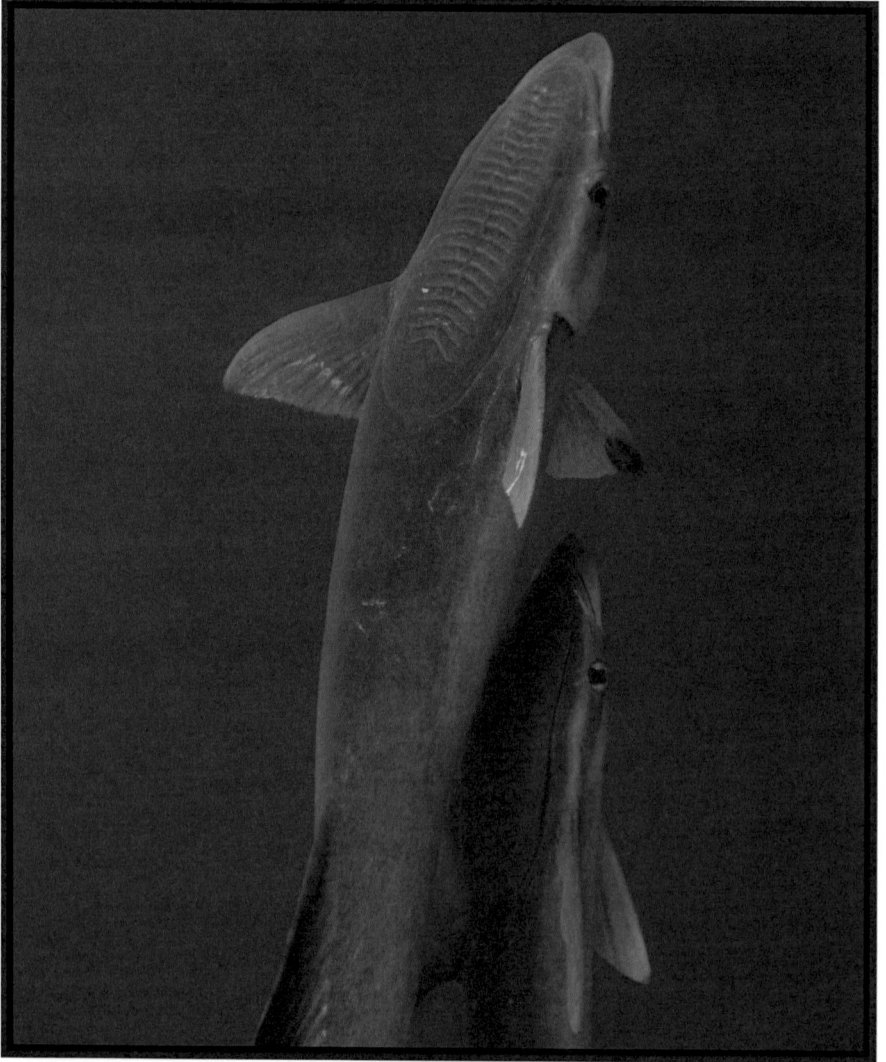

Remora

Echeneis naucrates

As a precautionary measure we often perform safety stops on our way back to the surface after exploring a reef site. Most of our stops are at a depth of 15 to 20 feet for a period of a few minutes in which sometimes we are visited by a pair of curious Remoras. The Remoras will swim around scoping us out as potential ride along hosts. They will even attach to a diver and/ or tank momentarily if they are really curious.

Given that Remoras are often referred to as "Shark Suckers," some divers tend to get a little nervous when these fish appear. Yes, Remoras are known to be associated with sharks, but they actually take advantage of nearly any larger animal in the ocean. This often includes turtles, manta rays, and large pelagic fish. The Remora has a modified dorsal fin that creates a suction cup type mechanism. The fish uses this suction to attach to larger creatures. The host animal benefits the Remora in the form of free transportation and leftover food bits from meals.

We encounter Remoras fairly routinely during our reef investigations offshore north Florida. If one should attach to you as a diver do not be startled as it will probably let go momentarily. It is not advised to try to pull it off as the suction can be a tremendous force. To give you an idea, Remora are reported to be used in fishing (in other countrys) for large animals such as turtles. A Remora can stay attached to a turtle well enough that when a fishing line is attached to that same Remora both the Remora and turtle can be dragged in by the fisherman.

Pictures captured at: "15) FF Concrete 2009" (page 42)

Sea Cucumbers

Multiple species

Sea Cucumbers are interesting animals that play a critical role in the marine ecosystem. They eat bits of organic particular matter found on the sea floor. The process involves ingesting significant amounts of sand and nonorganic particulates that the Sea Cucumber cannot digest. This explains why when we encounter these creatures we typically find a sand droppings path, in a tube-like shape, behind them.

Most often we find Sea Cucumbers in the open sand areas between reef structures but we also observe them on the encrusted reef structure as well. Several species are found offshore and below are two representations.

Top image:

Florida Sea Cucumber

Possibly: *Holothuria floridana*

Bottom image:

Chocolate Chip Cucumber

Isostichopus badionotus

Top picture captured at: "30) Little Barge Reef" (page 76)
Bottom picture captured at: "32) Matanzas Barge" (page 80)

Sea Slug

Hypselodoris picta or edenticulata

This nudibranch (sea slug) is often referred to as the "Florida Regal Goddess." There is some question in the taxonomy world whether this sea slug is its own direct species (*H. edenticulata*) or if it is a regional color variation of the western Atlantic species (*H. picta*).

The sea slugs I have encountered locally have all been very small and difficult to find. You might assume their intense contrasting color pattern would have them sticking out obnoxiously, but the majority we see are under one inch in length. Intense colors are not as dramatic on tiny objects to the human eye. The specimen captured in the photo to the left is the largest I have encountered and it was less than two inches in length. This nudibranch is often found on or near encrusting sponge mats which are presumed to be its food source.

Picture captured at: "10) Gibbs Dry Dock" (page 32)

Sea Urchins
Multiple species

All marine organisms mentioned in this book play an important role in the reef ecosystem and the sea urchin is no exception. Not only are urchins a popular food item for several carnivorous species, but they also serve in the maintenance of the reef structure itself. Urchins graze algae on the reef much like cows graze grass on farmland. Many algae species grow at faster rates than other encrusting organisms such as coral and sponge. These invertebrates (corals and sponges) compete for real estate with the algae and if left unchecked algae can out compete them. We have observed the evidence of urchin grazing with the placement of new artificial reefs. The first noticeable growth on the submerged structure is fast growing algae species. The algae grow rapidly and in a short period of time these new reefs become flooded with sea urchins. The urchins graze the algae down and other slower growing inverts begin to appear.

Top image:

Purple Sea Urchin

Arbacia punctulata

Bottom image:

Variegated Sea Urchin

Lytechinus williamsi

This urchin is often found in a white color with random shells adhered to its spines. The urchin will actually collect the shell bits to wear as décor to help blend into the reef surroundings as much as possible.

Sergeant Major
Abudefduf saxatilis

The Sergeant Major damsel is a very prominent damsel encountered at popular reefs in the Florida Keys. This brightly colored fish has come accustomed to being fed and is the first to greet snorkelers and divers in hopes for a free meal. These same fish are also common in north Florida waters and are found in the aquarium trade.

In my experiences, I have encountered most Sergeant Major from the shore. In fact, when I was a student at the University of North Florida I spent a great deal of time collecting juvenile specimens from submerged rocks near the Matanzas Inlet. Science partner Scott Dickie and I could see the bright colors of the little fish from the surface and were able to scoop them up with large dip nets. During the spring and summer months there were plenty of Sergeant Major near these rocky areas. I also find larger versions of the damsels offshore occasionally. Usually just a sole fish or a few are spotted, mainly at shallower reefs closer to shore. I presume they are more numerous on near-shore reefs and coastal areas with jetties, piers, and bridges.

Picture captured at: "7) Gulf America Wreck" (page 26)

Sheepshead

Archosargus probatocephalus

The Sheepshead is very familiar to northeast Florida recreational fishermen. The edible fish can be found both offshore and inshore making the species easy to access from the shore, piers, jetties, and boats of all sizes. Each year the Jacksonville Offshore Sports Fishing Club host a Sheepshead fishing tournament called the "El Cheapo." This has become one of the area's most popular fishing tournaments and it continues to generate interest in Sheepshead.

Sheepshead are found at wrecks offshore. In our experience we often encounter more at the reefs closer to shore. They seem to be unconcerned with the presence of divers, acting as if we are not even there as they go about their business. They closely scrutinize the reef looking for algae and invertebrates to snack on. Their mouths are equipped to crush barnacles, oysters, and other hard shelled creatures.

Pictures captured at: "7) Gulf America Wreck" (page 26)

Slippery Dick Wrasse
Halichoeres bivittatus

Like many wrasse species the Slippery Dick Wrasse has different color variations throughout various stages of its life. The pictures to the right, captured at reef sites offshore Jacksonville, show two variations.

Slippery Dick Wrasse are a common sight at most reefs in the area. We find they excitedly swim about the reef as close to the structure as possible. They are very difficult to capture pictures of because wrasses dart around in fast and unpredictable ways. The images provided involved several attempts and some luck.

Pictures captured at: "11) Press Box Ledge Reef" (page 34)

Sponge
Multiple species

Often when people think of reefs, visions of colorful stony coral gardens come to mind. While this perception is quite accurate for coral reefs found in the shallow seas of tropical regions, it does not quite match up with the reefs of north Florida. North Florida reefs are often colorful and garden-like but lack the great diversity of stony corals inhabiting tropical reefs. The offshore waters of Jacksonville and St. Augustine do include a few varieties of stony coral, but these corals are typically not vibrant and colorful. The color on these reefs comes from other encrusting invertebrates including sponges.

Sponges are found everywhere there is solid structure offshore north Florida. We encounter a great variety of sponges on nearly every reef regardless if it is a sunken vessel, concrete reef, or area of natural live bottom. They come in a variety shapes, sizes, and colors. We find barrel, branching, and mat forms with common colors of pink, white, orange, and yellow. Sponges are very much live animals. They are highly porous allowing water to circulate through their bodies. The water flow provides the circulatory, respiratory, digestive, and excretory needs of the sponge.

Sponges play a significant role on the ecology of north Florida reefs, providing both food and shelter for many other forms of marine life.

Picture captured at: "9) School Bus Barge" on page 30

Spotted Scorpionfish
Scorpaena plumieri

Unlike the invasive Lionfish, the Spotted Scorpion-fish is the native species of the family Scorpaenidae residing in the western Atlantic. As the name "Scorpion" applies, Spotted Scorpionfish have venomous spines. These fish camouflage very well, sitting motionlessly waiting to ambush prey that get close enough.

Ironically, my closest dive partners and I have only visually confirmed sighting the Spotted Scorpionfish at one location offshore St. Augustine. The site is a military aircraft wreck which is now serving as a small reef. At this site we found several Spotted Scorpionfish blending in with the growth on structural components of the remaining aircraft. My guess is these fish are so well camouflaged that we have not been able to find them on other reef sites. The aircraft wreck is a much smaller reef with less detail for our eyes to absorb. This may have made it more likely for us to find the Spotted Scorpionfish. At a typical large artificial reef there is an overwhelming amount of visual detail that may hinder our ability to isolate these extremely well hidden creatures.

Picture captured at: "St. Augustine plane wreck" (not featured)

Spotfin Butterflyfish

Chaetodon ocellatus

If you are familiar with the saltwater aquarium hobby or have seen such tanks on display, you are probably aware of the Spotfin Butterflyfish. The fact it is an attractive fish and considered a hardy specimen makes it common in the aquarium industry. The Spotfin is a butterflyfish found throughout the western Atlantic including the reefs of north Florida.

We regularly encounter isolated (or occasionally a few) pairs of Spotfin Butterflyfish during reef investigations. They are dramatically colored compared to most other fish and the reef background. This makes them easy to spot even though they are only a couple inches in length. The fish are observed combing the reef in search of tube worms and sea anemones to consume.

Picture captured at: "30) Little Barge Reef" (page 76)

Squirrelfish

Holocentrus adescensionis

Although the Squirrelfish is a nocturnal animal, we find them quite frequently during our day time diving expeditions offshore north Florida. They are rarely out in the open but usually tucked away deep inside dark sheltered areas. We typically see them while looking under ledges and overhangs for other creatures. Squirrelfish will become active at night to hunt for crustaceans. Their large eyes in proportion to body size adequately compliment the nighttime lifestyle.

The fish we encounter are usually in the four to six inch length range. We tend to observe them more frequently on sunken vessel reefs, but it is expected they utilize most reef habitats in the region. Reef base structures in themselves are probably not a significant factor in Squirrelfish populations. Concrete, metal, and natural stone reefs all provide food and refuge preferences of this big-eyed but smaller fish. So regardless of what type of reef you are exploring offshore Jacksonville or St. Augustine, make sure to bring a light to look into sheltered areas. Even if you do not encounter a Squirrelfish you are bound to find something interesting.

Picture captured at: "30) Little Barge Reef" (page 76)

Summer Flounder

Paralichthys dentatus

Summer Flounder is a favorite local food fish. Their meat is white and light with absolutely no fishy taste. Fortunately for anglers and sea food enthusiasts, Summer Flounder are common in northeast Florida waters.

Summer Flounder are left side flounder meaning their two eyes are on their left side. They lack a swim bladder and have pigmentation that can adjust to their environment. These two traits allow flounder to rest on the sea floor in disguise, making it very difficult for a diver or a prey species to locate them. In fact, I often encounter Summer Flounder by accident with my camera tripod. There have been many dives where I have set my camera tripod leg down on a hidden Summer Flounder while setting up for a shot of something unrelated. Of course at that point the annoyed fish will swim away allowing me to see it. On some reef investigations Summer Flounder have been so abundant that I would repeat this scenario three or four times a dive. Each time I would place my tripod legs directly on the sandy sea floor only to be startled to watch a flounder dart away.

I would say we observe Summer Flounder most often on concrete material artificial reefs. The fish are commonly found hidden in the sandy open spaces between structures. Summer Flounder have a buffet of food items available to them at these sites, and simply have to wait inconspicuously until the meal of choice swims by.

Top left picture captured at: "15) FF Concrete 2009" (page 42)
Top right picture captured at: "1) MR East Barge" (page 14)
Bottom picture captured at: "St. Augustine planewreck" (not featured)

Tomtate Grunt

Haemulon aurolineatum

If I had to pick a fish species and claim it to be the most abundant fish routinely encountered on north Florida reefs, I would choose the Tomtate Grunt. Of course there are other popular fish such as Black Sea Bass and Cigar Minnows but Tomtates outnumber resident Black Sea Bass populations and minnow schools are not consistently observed. This is not a fish targeted by fisherman, except for perhaps bait purpuses, but it is an important component to the marine food chain. As a small popular reef fish it is a common food source to many larger animals.

Tomtates form thick mats of tightly swimming fish over and within reef structure. At times the congregations are dense enough so that the benthic habitat is completely covered. More often than not when we are surveying a sunken vessel reef, the open space within the vessel will be so crammed with Tomtates that we are unable to see anything but a wall of fish.

Pictures captured at: "17) Coppedge Tug" (page 46)

Tube Worms

Multiple species

There is a seemingly endless variety of interesting tube-type worms found offshore. Most varieties have an exposed feathery or tentacle fitted end that can completely retract when the animal is startled. These exposed portions of the worms often make for a very attractive display.

Top image:

Anemone Worm

Genus species?

Middle image:

Christmas Tree Worm

Spirobranchus giganteus

Bottom image:

Feather Duster Worm

Possibly: *Sabellastarte magnifica*

Twospot Cardinalfish
Apogon maculatus

These colorful little fish are occasionally found in the aquarium trade where they are often referred to as Flamefish. They reportedly inhabit broad ranges of the western Atlantic Ocean and the Gulf of Mexico. We routinely encounter these nocturnal creatures during our daytime diving investigations tucked away in sheltered dark areas within reef structure.

The average Twospot Cardinalfish we observe offshore northeast Florida is approximately two inches in length, bright red in color, and has two dark spots. I personally encounter specimens much smaller then this at times. These smaller Twospot Cardinalfish are often so small they are difficult to identify and are usually found in small groups. The larger adults are generally found in pairs, as solitary individuals, or in small aggregations. Twospot Cardinalfish are encountered at nearly every reef site we survey here in north Florida.

Picture captured at: "18) Coppedge Culverts" (page 48)

White Spotted Soapfish
Rypticus maculatus

This fish, although generally less than eight inches in length, is in the same family as grouper (Serranidae). It is referred to as the Soapfish because its skin releases a toxin called grammistin. The soapy slime probably tastes bad and thus helps the fish avoid predation.

Soapfish are extremely common on north Florida reef systems. We encounter them during every reef dive in substantial numbers. Usually they are tucked away in structure sticking only their heads out just enough for us to see their eyes. They almost look like sad puppy dogs looking up at you.

Pictures captured at: "10) Gibbs Dry Dock" (page 32)

Invasive Lionfish

Most of us are aware that our world's oceans are bombarded with forces negating the health of marine ecosystems. Sadly many of these attacks, such as pollution and over-harvesting, are caused by humans. It is a shame that we can abuse the same system that provides so much for us.

The marine environment off northeast Florida is not invincible to these negative pressures. In fact, there are many ailments affecting our water quality and the life it supports. Several books would need to be written to discuss each issue, however I want to briefly mention one. I feel our community is not fully aware how detrimental this problem is.

I am referring to an invasive species problem plaguing the coastal offshore waters of the entire western Atlantic. The culprit is known as the Lionfish, which is native to the Indo-Pacific region. Since the early 2000's Lionfish populations in the western Atlantic have been exploding. Today heavy Lionfish populations are found the along the entire east coast of the US, throughout the Gulf of Mexico, and along the east coast of South America. While investigating reefs offshore northeast Florida, we have encountered over 100 Lionfish per reef location on several occasions. Ironically, in their native home range, Lionfish populations are mysteriously kept in check and it is actually somewhat rare to encounter one.

There does not appear to be a natural checks and balances

system to control Lionfish populations here. They effectively have no predators and prey on basically anything they can fit in their mouths. Local Lionfish are feared to consume more and breed more frequently than many native fish species. This creates many challenges for our native species. Not only do they risk being consumed but they also have to compete for food with the invaders. These conflicts span across several layers of the marine food chain affecting recreational and commercial targeted fish species. Lionfish have been found with juvenile grouper and snapper in their guts. Lionfish are voracious and indiscriminate. They do not care about endangered or threatened animal species, nor do they abide by regulations in place to protect those creatures. They eat what often includes species of ecological and econmical importance in addition to the food sources that support those species.

Unfortunately there is no easy quick fix to deal with this issue. We, as concerned stewards of our environment, need to do what we can to responsibly remove as many Lionfish from our local waters as possible. The organization REEF has been a pioneer in this movement. REEF encourages scuba divers to harvest lionfish and consider eating them. The meat, a lean protein source, is said to be very tasty by those who have tried it. If enough Lionfish harvesting efforts can be implemented around the invaded areas at frequent rates, there is hope we can start to control this dangerous and invasive predator.

Disclaimer:

Lionfish have venomous spines within their dorsal, pelvic, and anal fins. If any of these spines are to pierce human flesh the resulting injury may be extremely painful and perhaps more severe. It is strongly encouraged those wishing to harvest Lionfish reference other sources pertaining to the proper handling of Lionfish. The REEF organization is a great resource for such information. With a few precautions in mind informed divers can be on their way to helping combat the Lionfish invasion.

TISIRI Corp is a 501(c)(3) non-profit organization focusing its efforts in marine habitat creation, environmental conservation, awareness, and underwater exploration. TISIRI stands for "Think It, Sink It, Reef It."

Artificial reef creation is an effective process to develop productive marine habitat. Placed reefs become offshore destinations benefiting the environment and surrounding communities. TISIRI generates reef concepts incorporating recycled material, works with regulatory agencies to secure/confirm proper permitting, and solicits private funding to create marine bio-diverse hot spots. These reefs are appreciated by fisherman, scuba divers, environmentalists, industries, and marine life.

Please visit the TISIRI website to learn more at www.TISIRI.org.

www.ingramcontent.com/pod-product-compliance
Lightning Source LLC
Chambersburg PA
CBRC090214270326
41929CB00006B/36